THE MAN WHO
TASTED WORDS

ALSO BY GUY LESCHZINER

The Nocturnal Brain

THE MAN WHO TASTED WORDS

A Neurologist Explores the Strange and
Startling World of Our Senses

GUY LESCHZINER

ST. MARTIN'S PRESS
NEW YORK

First published in the United States by St. Martin's Press, an imprint of St. Martin's Publishing Group

THE MAN WHO TASTED WORDS. Copyright © 2022 by Guy Leschziner. All rights reserved. Printed in the United States of America. For information, address St. Martin's Publishing Group, 120 Broadway, New York, NY 10271.

www.stmartins.com

Images pp. 42, 71, 96, 204, 249 © Jill Tytherleigh

Images p.79 (top) Granger Historical Picture Archive / Alamy Stock Photo;
(bottom) M.C. Escher's "Ascending and Descending" © 2021 The M.C. Escher Company – The Netherlands. All rights reserved. www.mcescher.com

Library of Congress Cataloging-in-Publication Data

Names: Leschziner, Guy, author.
Title: The man who tasted words : a neurologist explores the strange
    and startling world of our senses / Guy Leschziner.
Description: First U.S. edition. | New York : St. Martin's Press, [2022] |
    Includes bibliographical references and index.
Identifiers: LCCN 2021046504 | ISBN 9781250272362 (hardcover) |
    ISBN 9781250272379 (ebook)
Subjects: LCSH: Senses and sensation. | Sensory disorders. | Brain.
Classification: LCC QP431 .L47 2022 | DDC 612.8—dc23/eng/20211108
LC record available at https://lccn.loc.gov/2021046504

Our books may be purchased in bulk for promotional, educational, or business use. Please contact your local bookseller or the Macmillan Corporate and Premium Sales Department at 1-800-221-7945, extension 5442, or by email at MacmillanSpecialMarkets@macmillan.com.

Originally published in Great Britain by Simon & Schuster UK Ltd

First U.S. Edition: 2022

10 9 8 7 6 5 4 3 2 1

For Frida and Michael

CONTENTS

INTRODUCTION

'Man has no Body distinct from his Soul for that call'd Body is a portion of Soul discerned by the five Senses, the chief inlets of Soul in this age.'

William Blake, *The Marriage of Heaven and Hell*

'And God said, "Let there be light", and there was light.' And there was the sound of running water, the feel of the breeze on Adam's face, the smell of the flowers, and the taste of the apple in Eve's mouth. Thus was the world born, and thus are we born into this world, from the moment our eyes open, dazzled by the light, our nostrils filled with the scent of our mother, the sweetness of milk on our tongue, the soothing sound of the maternal voice and the warm, comforting feel of skin on skin. The universe is brought into sharp reality as we begin to perceive our surroundings through our senses. And, indeed, we are reborn into this world with every waking moment, the transition between our dreams and the cold, hard world played out as we crack open our eyes, the morning sounds of the hum of traffic or birdsong drawing us down from our slumber, back to Earth with a bump.

Consider any instant in your life, from the mundanities of your daily grind to those special, treasured moments – the smell of the nape of a loved one's neck, or that of a freshly brewed cup of coffee; the taste of a dish that takes you straight back to childhood – a fragmentary, comfortable memory of happy times; your favourite track suddenly playing on the radio; the familiar sight of a display board on the train platform, signalling the delay of your morning train; the feel of your child's hand in your own.

These snapshots of our lives constitute the merging of our external and internal worlds, the coming together of our memories, emotions, histories and desires, and our environment. And it is upon our senses – vision, hearing, taste, smell and touch – that we rely to perceive the reality of our world beyond our own body. These senses are our windows on reality, the conduits between our internal and external lives. They are how we absorb the outside world. Without them, we are cut off, isolated, adrift. We cannot live anything other than a virtual life, within the realms of our own minds.

My earliest memory is orange. Not the fruit, but a lurid, acid-tinged colour peculiar to, and the quintessence of, the 1970s. I can see the sky above, but all around me, on every side, is orange.

For many years, I could never quite place this memory of orange. Its origins remained a mystery, its age unknown. It is some years later, perhaps in my teens, that I come across a photo in a family album; vintage tones and curled at the edges. My mother, her hair tightly curled, stands in the centre of a square in the small village in West Germany of my early childhood. Next to her, a pram; a small infant, me, inside it. The pram has the sheen of plastic, a vinyl that must have been the height of modernity in the mid-'70s. And the colour of the pram is orange, the precise shade in my mind's eye. All of a sudden, I understand. The blurry vision of the past: me, sitting in the pram, looking up, the hood of orange fabric framing the square of clouds and sky.

But another explanation occurs to me. Perhaps I have seen this photo before, in the months or years after my family came to England, looking through the scanty souvenirs of our past life in another country. And perhaps I have seen this pram, this colour, many times before. Maybe this memory that I have – that I have always held to be the earliest conscious remnant of my life – is not real. Maybe it is a false memory that I have created, a fictional portrayal of my past, a betrayal by my mind.

We are all familiar with the concept of inaccurate or incomplete memory, aware that our recollection of events may fade or degrade with time. We may misremember or forget entirely. We may even create memories from nowhere. The shortcomings of our brains are readily apparent. But perhaps there is another possibility. It is not just the memories of our experiences that are vulnerable to the vagaries of the functioning of our brains. It is the experiences themselves.

The sights, sounds, smells, tastes and feel of the world around us are solid, crisp, distinct and real. We do not doubt them; 'Seeing is believing.' The act of sensing something for ourselves cements it in reality – no longer a story told or heard, experienced second-hand, but a fixed representation of the real world around us, as solid as the floor beneath our feet, and sharp as the knife blade that cuts our finger, as bright as the sun blinding us with its light. Our sensations are our portal into the physical world that envelops us, moulds and shapes us. Through the act of sensation, doubt is cast aside, our faith in what we see or hear more absolute than a devout person's belief in God. In

Aristotle's view, the five senses are the basis of all knowledge, through which we observe the 'essence' of the world; through our senses, the material world interacts with our psyche. Our inner world, our mind, is like soft wax imprinted with our sensory experiences.

Yet perhaps we should be more agnostic, less fervent in our trust of our senses, more questioning of our eyes, ears, skin, tongue and nose. We imagine that these organs that convey our senses are reliable and precise witnesses of our external worlds, accurately reporting on the colour of the bunch of roses we gaze upon, or the pain of pricking our finger on one of the thorns. But we imagine wrongly. What we believe to be a precise representation of the world around us is nothing more than an illusion, layer upon layer of processing of sensory information, and the interpretation of that information according to our expectations. Like the shaded shape on a flat piece of paper that we see as a three-dimensional object, or the feel of an itch without any obvious cause. What we perceive to be the absolute truth of the world around us is a complex reconstruction, a virtual reality recreated by the machinations of our minds and our nervous systems. And, for the most part, we are totally unaware of these processes; we are shaken by the moments when the discordance between our perception and reality are revealed, such as gazing at a drawing by M. C. Escher or squabbling over whether a dress is white and gold or black and blue.

Our sensory end-organs – our eyes, ears, skin, tongue and nose – are merely the first step in these pathways of perception; what we then experience as, for example, vision or sound relates only loosely to the beams of light falling onto our retina or the sound waves vibrating tiny hair cells in the cochlea of the inner ear. Moving beyond the point at which our bodies physically interact with the world, the complexities of our nervous systems come into play, acting like supercomputers, fundamentally altering what it is we actually feel, taste, smell, see or hear. The translation of these basic inputs into experiences with conscious meaning – the pattern of light and dark on the retina transformed into the face of a loved one, or the feel of a cold, damp object in the hand and bubbles gently fizzing on the tongue experienced as a delicious glass of champagne – is a process of utmost abstraction, simplification and integration, invisible and undetected by us. The pathways, from physical environment to our experience of it, are convoluted and complex, vulnerable to the nature of the system, friable in the face of disease or dysfunction.

In the pages that follow, I will introduce you to a range of individuals whose senses have been altered or changed in some way, imbued with a diminishment or amplification of their perception of a particular aspect of their world; whose version of reality is shaped by their senses in unusual and often dramatic ways. For some, their condition has been present from birth; for others, it was acquired in later life. For many, their experience is deemed a 'disease' or 'disorder', but there are other persons who sit within the spectrum of normality for humanity, despite the almost implausible world they inhabit. For all of them, the nature of that difference has been transformative, in some cases making life as we know it unrecognisable. Some of these people are my patients; some are from other walks of life. They are all extraordinary, not only because of what they have experienced, but because they have been kind enough to share their stories. As ever in the world of neurology, it is through understanding the system when it goes wrong that we gain insight into normal function.

The stories relayed in this book starkly illustrate the

limitations and idiosyncrasies of our senses, for every single one of us – their reliance on the structural and functional integrity of our nervous systems, and, importantly, how the perception of our world, for all of us, may be rather different from the reality. Their experiences raise questions about the very nature of reality, and what it is to be human.

I

THE STUFF OF SUPERHEROES

'Of pain you could wish only one thing: that it should stop. Nothing in the world was so bad as physical pain.'

George Orwell, *Nineteen Eighty-Four*

'An ordinary hand – just lonely for something to touch that touches back.'

Anne Sexton, 'The Touch'

'When one of my teeth fell out as a child, my dad made the mistake of telling me that, if I put it under my pillow, the tooth fairy would give me a pound,' Paul, now thirty-four, tells me. 'I immediately thought to myself, 'Oh, great! Well, I've got many teeth in my head. That's many pounds!' he chuckles. 'My dad caught me with a pair of pliers, trying to pull my teeth out.' As I sit at the kitchen table with Paul while his father, Bob, and his mother, Christine, potter around behind us, the horrifying stories of Paul's childhood pour out. Paul turns to his parents and says, 'I remember once asking for snacks or crisps, and you said I couldn't have them because dinner was coming soon. And I just stood there, breaking my own fingers because I couldn't get what I wanted.' He mimes bending back his own fingers, and, in my head, I can hear them cracking. 'Yeah, I did stupid things. Things that, obviously, any normal child wouldn't dream of.' And it is obvious, when I listen to Paul and Bob talk, that Paul was anything but a normal child. Indeed, he is not a normal adult – because Paul feels no pain. None at all. He has never felt it. He has no concept of what pain is. He tells me, 'I have a hard time showing empathy to someone who is in pain. It is hard understanding pain when you don't feel it yourself.'

The inability to feel pain is the stuff of superheroes, the deepest wish of those tortured by it. But Paul's inability to sense pain is unfortunately not coupled with super-strength, unbreakable bones and super-healing. I ask Paul to estimate how many times he has broken a bone. 'It's got to be in the hundreds, from minor fractures to major breaks. Fingers, ankles, wrists, elbows, knees, thighs, skull – I don't think there is a bone I haven't actually broken.' When I had first come into the house, Paul was already seated at the table. My first impression was of a young man with fair hair, bespectacled – ordinary. If I passed him in the street, I would not give him a second glance. As we chat, I can see that his hands are a little misshapen, but it is not until I go to leave, and he stands up, that I see how short he is. He is perhaps just over five feet tall. 'The only reason I am as short as I am is because of the damage I did to my knees as a child. I broke the growth plates in my knees on multiple occasions, which stunted my growth significantly.' And when he walks towards the door as I leave, I see his limping gait, his bowed legs, evidence of endless badly healed fractures.

Consider your five senses. Place them in order – at the top put the sense you could not survive without, at the bottom the one you would sacrifice first. A fantasy football league; a ranking of players you need to win the tournament. For me, vision would definitely be at the top. Losing my vision, unable to read, to see the faces of friends and family, to look at a beautiful landscape, would be too much to bear. Then hearing: not to be able to hear music or speech would be almost as intolerable. Both these senses allow us to detect the world at a distance, to know the environment beyond the immediacy of our bodies, to derive pleasure, detect warning, to engage in societal interaction and the exchange of ideas and concepts. At the bottom of my list, in line for relegation, would be smell and taste. Horrible to be without the rich world of food, or devoid of olfactory pleasure, but my life would go on. Touch – well, it wouldn't really compete with vision or hearing, so it sits in position number three.

But take a step back for a moment. Consider a life without touch: the inability to feel the hug of our loved ones, the warmth of the sun on our face, the warning of heat as we approach a fire. However, touch is more than just these sensations. We rely on our touch to walk properly, to feel the undulations and irregularities of the floor under our feet, to know where our bodies are in space, how the position of one hand relates to the other while tying our shoelaces or eating with a knife and fork, to pull the right coin out of our pocket when paying a bus fare. Without touch, even these very simple acts would be impossible. While one imagines touch to be a lesser sense, perhaps the very opposite is true. Perhaps touch is so intrinsic to the act of being, so enmeshed in our existence and our consciousness, that it is almost impossible to imagine a life without this sense. Our language reflects this to a great extent. We describe people as 'warm' or 'cold', 'soft' or 'hard', ascribing character or feelings to physical sensations. We use phrases like 'I am touched by your kindness', 'She's a pain in the arse', or 'He can be hot-headed'. The language of life is based much more upon touch than on hearing or vision. But these are not just linguistic patterns. This translates to reality. Experiments have shown that, depending on whether you clutch a hot drink or a cold one in your hands while talking to someone, you will perceive your conversation partner as 'warmer' or 'colder', and that handling a block of hard wood or a soft piece of material while interviewing someone will influence your perception of them respectively. The warmth of resting on our mother's chest, the association with a feeling of safety and comfort, pervades the rest of our lives – an intrinsic aspect of human nature and linguistics. Touch links us to those around us, the binding effects of a hug, a touch on the arm, a pat on the back, a caress. Our sense of touch goes far beyond simple electrical impulses triggered in our skin, but is entwined with our emotions, memories, sense of self and sense of others. And seeing the impact of disruption of this sense in many of my patients, I would now certainly not volunteer to lose touch above other senses.

As you will discover over the course of this book, the absence of sensation can be devastating. But an absence of pain – the loudest of our sensations – sounds like a blessing, not a curse. Pain screams its way into our consciousness, blotting out everything else. The blinding sear of stubbing one's toe, cracking one's head, or cutting your finger, elbows all other sensations and senses out of the way, demanding immediate attention and action – and, as Paul demonstrates, for good reason. Pain prevents us from injuring ourselves, or at least from making the same mistake twice. We need pain to help us learn to avoid sharp or hot objects, to teach us what in our environment is potentially harmful, and to detect injury or infection. If we do injure ourselves, pain focuses our attention on looking after that part of the body, protecting and immobilising it so that we can repair and heal before we start using it again.

These multiple functions of pain are reflected by its various qualities. One important aspect is knowing where the pain is coming from, localising the site of injury or damage. It is crucial to our survival to know that the agony we are feeling is due to having burned our finger on a hot pan, or from a thorn in our left big toe.

But pain also has an emotional component – that gut-wrenching unpleasantness, that fear – that is a potent driver of learning to avoid it. Without the emotional baggage that accompanies that sensation of hurt, we would be less inclined to learn from our mistakes, to develop strategies to prevent repeat incidents. The risks would be too great, our lives curtailed, the survival of our species jeopardised. In fact, our brains are evidence of the significance of the emotional aspects of pain to our evolution. The areas of the brain involved in this aspect of our experience of pain are in the oldest evolutionary parts of our brains, structures that developed millions of years ago in the evolutionary pathway of animals, preserved in per-petuity, the signature of the utility of pain.

Studies in animals and humans show that multiple areas of the brain are involved in the perception of pain. There is not one single spot, no single area of the brain, where pain is 'felt'. In fact, the underlying brain mechanisms of pain perception are more like a network rather than a single pathway. This network reflects our understanding of the different aspects of pain: the ability to identify where in the body pain is, termed the 'sensory-discriminative' component, and the emotional load, often referred to as the 'affective' component. Separate but interrelated.

Information about where the pain is coming from is relayed to the area of the brain involved in all aspects of touch – the sensory cortex. This strip of brain tissue is the location of the homunculus, the brain's sensory map of the body. When represented in a diagram or model, it shows a grossly distorted figure with overblown lips, tongue, hands and feet, where the density of our sensation receptors is highest and the requirement to discriminate the precise location of any touch is most pronounced. Simultaneously, this information about pain is relayed to even more evolutionarily ancient areas of the brain – those responsible for our emotions and drives; regions of the brain that encode our primitive needs – beneficial ones such as hunger, thirst and sexual desire – and those that are aversive – such as fear, danger and, importantly, pain. And it is here, in the limbic system, the emotional nexus of the brain that resides in the central depths, that the affective component of pain is processed.

One area of the limbic system in particular, the anterior cingulate cortex, is implicated in the unpleasantness and fear of pain, and is a potent driver of the need to avoid pain. Damage to this area of the brain results in a phenomenon termed 'pain asymbolia', which is the perception of the precise location, quality and intensity of pain without the emotional context, leaving individuals indifferent to pain and slower to withdraw from it, due to the lack of an emotional driver to avoid it again at all costs. Similarly, destruction of pathways leading to the area of the brain responsible for our body map can result in people experiencing the negative emotional impact of pain without knowing where it is coming from.

I remember my own children as toddlers – a slip down a few stairs or the sting of a fall an important reminder to watch their step. A smack from their sister: a lesson in how to respect their sibling's toys. One of my own earliest memories is from the age of three or four. I remember a hot, sunny day, typical for summer in a small village on the fringes of the Black Forest, a few miles from the Rhine, defining the border with France. I had been playing with my friends, cycling around, having fun in the playground, the air filled with excited shrieks of joy. We were like a gang of street urchins, roaming the streets of the village, uninhibited by adult supervision. The heat of the sun was waning, and I recall being tired and hungry, the large, heavy glass door to our block of flats a hurdle between me and my dinner. As I heaved the door open, it struck a bee, which promptly flew at me and stung me on the arm. I can still see the pulsing sac of pain-inducing venom pumping into my skin. Those shrieks of joy were quickly replaced by my howls of agony – and I developed a newfound respect for all things flying and stinging.

But for Paul, these life lessons are entirely alien concepts. As a child there was no clue for him not to do certain things. If anything, Paul sought reward through damaging his own body. 'I used to do stupid things like jumping down a staircase, or jumping off the roof. There was no downside for me. I didn't feel any pain. All I saw was everyone around me giving me loads of attention.' He recalls spells in hospital, surrounded by doctors and nurses, feeling spoilt and fussed over. Perversely, for Paul, injuring himself was a positive experience. His father, Bob, recalls one episode when he found his son standing on the flat roof of the garage. 'I panicked. And the next-door neighbour said, "You see, Bob, there's your problem. You know that children play to the gallery. You should say to him, 'If you want to jump, Paul, go ahead and break both your legs.' It's reverse psychology!" I said, "Let's see. I think you're right!" So I said, "Paul, if you want to jump and break both your legs, and spend the next two months in hospital, it's completely up to you." And straight away, he jumped off the roof and broke both his legs – and spent several weeks in hospital. He loved it.'

The reason for Paul's complete inability to feel pain is an

extremely rare genetic condition called congenital insensitivity to pain, or CIP. Since the moment of his birth, he has never experienced feeling physical hurt – no headache, toothache, or any other ache. Bob reports that Christine, Paul's mother, was aware of something odd about Paul right from the start. He remembers her saying, 'Don't you think it strange that he never cries?' Bob just assumed that Paul was a happy baby. But one day, when Paul was about ten months old, he was lying on the floor, surrounded by cuddly toys, when Bob came in from work. Bob recalls: 'Suddenly Christine jumped, because I was standing on Paul's arm! I hadn't realised because of all the toys all over the floor.' Despite an adult standing on him, Paul still didn't cry. Not a peep.

By this time, Christine was convinced that Paul was very different from other babies. It wasn't until some time after this incident, when Paul developed some sort of abscess and was taken into hospital, that his condition came to medical attention. The doctor asked if Paul had been crying in discomfort, and Bob told him, 'My wife has got the crazy idea that he doesn't feel pain.' And so began the process of getting Paul diagnosed. Bob tells me, 'We went to Great Ormond Street Hospital and they put these electrodes on him. They said, "We'll go up 10 volts at a time. He will feel pain in some part of the body." They got quite upset because the veins in his face and arms bulged, but they went up to 300 volts and couldn't find any pain reaction in any part of his body. And I remember saying that it would be nice if he grows up to be a boxer, but of course I didn't realise the implications of not feeling pain.'

I am curious as to whether Paul's understanding of psychological pain is also affected; whether the absence of physical pain has somehow hindered the development of those parts of the nervous system that process emotional angst. Has he experienced the pain of heartbreak, the ache of loss? But as far as he can tell, this aspect of his life is the same as everyone else's. 'I've been told on many occasions, growing up, that emotion and [physical] pain are all linked,' Paul tells me. 'I feel touch, I feel emotion, I feel all the other senses. I feel them all except for pain.' I ask him if he understands on a personal level, or a purely intellectual one, when people talk about the pain of a broken heart or the pain of sadness; whether the empathy he is unable to feel when he sees people in physical pain extends to emotional hurt. But Paul is very clear on this. He has lost several people in his life, family members who have died. The internal ache, the deep and gnawing sense of loss, is something that he is sadly very familiar with. And as we discuss life more broadly, it is obvious that he feels the pain of lost opportunities, unrequited love, unfulfilled dreams. In Paul, there is a disconnect between physical and emotional pain. At first glance, this seems somewhat unfortunate – if you are unable to feel physical pain, perhaps it would be better to lose pain altogether. But without the distress of loss or the fear of it, perhaps there is also no joy from love, no ache from wanting. Without this emotional depth, what would our lives be? Like that of a psychopath, unable to form relationships, incapable of empathising with other people's lives.

The ability to feel this emotional pain implies that the central networks controlling this aspect of pain sensation are present in Paul, unaffected by his condition. His problem is more fundamental, simply concerning the perception of physical pain itself. Injury to his body and the normal triggers of tissue damage from burning, cutting or inflammation are just not making their way to the brain itself.

The conduction of impulses throughout the nervous system is dependent on a piece of very specific molecular machinery called the sodium channel. Sodium channels exist as molecular pores on the outer membrane of nerve cells – also known as neurones – like the holes of a fine sieve. In contrast to a sieve, however, for the most part these pores remain closed, and are triggered to open only under certain conditions. When triggered, the sodium channel opens, allowing sodium ions and their associated positive electrical charge to flood into the cell, like water exiting the bathtub when the plug is pulled. This small shift in electrical charge across the surface of the nerve cell does not itself result in the transmission of signals, but it is the nature of the trigger that opens the sodium channel that is key to this physiological process that is the keystone of life. The sodium channel has a very particular quality: it detects small changes in electrical charge, with even a small flow of ions in its vicinity causing it to open. And so, the opening of one sodium channel causes the ones next to it to open, thus generating a fall of dominoes and the rapid spread of this electrical impulse along the length of the nerve cell. A Mexican wave, each sodium channel like a spectator waiting for the fan next to them to stand up, conveying a message from one side of the football stadium to the other – or, in this case, from one end of a nerve cell to the other.

Sodium channels exist in a variety of forms, each with subtly different properties and each residing in different concentrations in various parts of the body. Some channels, rather than opening in response to changes in electrical state, are triggered by chemical transmitters, such as those responsible for muscle contraction. In this case, the electrical impulse travelling down the nerve cell causes the nerve endings to release a chemical called acetylcholine. Sodium channels in the muscle fibres sense the acetylcholine and open up, triggering a wholesale chemical response that results in movement. However, it is the sodium channels that are triggered by changes in electrical state that are primarily responsible for the sending of electrical impulses along our nerves.

Some types of sodium channel are more heavily implicated in the conduction of pain signals. Paul's condition clearly tells us that one particular sodium channel is crucial for the conduction of pain. Paul's problem is a change in a gene called SCN9A, the repository for the genetic information for a form of sodium channel called Nav1.7. These Nav1.7 channels are particularly concentrated in pain-transmitting pathways, and any change in their function seems to especially influence the processing of pain signals. Paul's Nav1.7 channels are entirely inactivated. The genetic error in Paul doesn't simply produce a Nav1.7 channel that's difficult to trigger, for instance; the mutation Paul carries is so catastrophic that no functional channels are produced at all.

For Paul's condition to manifest, however, it is not sufficient to have a single mutation. For almost all genes, we carry two copies, one inherited from our mother and the other from our father. So, if we inherit one copy of a gene that produces no channel at all, the second version of the gene should still deliver. While Bob and Christine are both carriers of Paul's condition, they are completely unaffected because they each only carry one abnormal gene. They were entirely unaware of this, at least until they started a family. But Paul inherited this abnormal gene from *both* of them, leaving him with no functional Nav1.7 channels anywhere in his body. The machinery fundamental to the transmission of pain impulses is entirely absent in Paul, leaving him completely devoid of pain.

Paul's very specific loss of molecular function results in a very specific physical dysfunction, while other molecular machinery is left untouched. I ask Paul if he can feel the heat of a spicy curry, or the cooling effects of menthol. While he can feel the heat of chilli, he says it is not unpleasant – there is no burn, no pain, no discomfort associated with it. He recalls being in a restaurant a few years ago with a friend. 'They had really hot chillies in this restaurant, and I dared my friend to be able to eat one. So he took one bite, and he was sweating and his mouth was on fire. But I sat there and ate five. I felt the heat in my mouth, but by no means was it uncomfortable or painful.' A rather unfair contest, I think to myself. While the temperature sensors in Paul's skin, including the sensitive skin of the mouth, are working normally, the associated pain signals have simply vanished – an important illustration of how various aspects of sensation are conveyed almost entirely separately, like parallel train lines carrying different types of passengers to the same destination. And, even more curiously, Paul discovered late in life that he has no sense of smell. In addition to its role in producing pain, the sodium channel he is missing has an important function for smell – a very odd combination of such specific roles.

Through further terrible misfortune, Paul is not the only member of the family affected. He is one of three siblings and, despite miniscule odds of probability, all of them have been affected. If you consider the statistics, it seems almost impossible. Each child born to Bob and Christine has a one-in-four chance of inheriting both abnormal genes, making the probability of three affected siblings one in sixty-four. And the probability of two people who carry an abnormal gene then meeting and having a family is vanishingly small in itself, which is why this condition is as rare as it is, affecting at most a few handfuls of people in the entire world. But Christine knew in her heart, even before Paul's younger sisters were born, that they too would have the condition.

The effects of this disorder have been devastating for the

entire family. For Christine and Bob, it has been awful beyond comprehension. Amanda, their youngest child, did not survive the absence of pain. At the age of thirteen months she succumbed to sepsis, unrecognised by doctors, in part due to the lack of warning signs one would expect to see from a child in pain. As if this weren't enough, ensuring that their surviving children, Paul and his sister Vicky, remained alive was almost an impossible task. 'As parents, we've been frightened every single day of our lives.'

As we sit at the dining table, I notice that in the corner of the room there is a stone-clad fireplace; it houses a gas fire behind a panel of glass that seals the flames from the room itself. Paul looks over to it and chuckles. 'One of our games was to hold our hands up against the glass when the fire was on. We used to love to hear the sound of the sizzling of our skin.' Bob shrugs: 'They would be laughing in the living room, listening to their hands sizzling like bacon in the pan, blisters all over.'

Bob also recalls the two siblings playing a game in the back garden, hearing them playing on the swings, their laughter filtering through the patio doors. 'Chris said to me, "Just check on them, will you?" And I said, "They're okay, Chris, they're just playing!" But when I went outside to look, Paul and Vicky were completely covered in blood. Vicky had knocked out all her teeth; Paul had knocked out his teeth.' Both the children needed stitches to their heads, and had black eyes and broken noses. The next day, a family outing to a show resulted in unexpected consequences. The sight of two children, bandaged and bruised, caused quite a stir. By the time the family got home, the police were waiting for them. Bob, with some bitterness, remembers being taken into the dining room by one of the policemen. 'He said, "Do you think it's clever for a seventeen-stone man to beat up a child?" And I won't tell you what he called me.' Despite Bob's protestations, he was threatened with arrest, until the police learned the medical background. An apology soon followed, as did a collection for the children from the local police station.

Over the years, Bob and Christine had various interactions with social services, and several threats to have the children removed. For them, the lack of understanding of their children's condition has been almost as distressing as the lack of pain itself. But perhaps this is not surprising. For almost all of us, pain is intrinsic to the human experience. We are aware of it from our earliest consciousness, and the language of pain is interwoven with our lives, telling us what we can and cannot do, affecting our day-to-day existence. There are medical specialties entirely dedicated to its control and elimination. So while the absence of it is, intellectually, perhaps just about comprehensible, on an emotional level it is totally beyond our understanding. And its rarity means that even most healthcare professionals are entirely unfamiliar with it. Paul describes one episode, a few years ago, when he woke up in the middle of the night. As he sat up in bed, he felt a crunch in his leg, a vibration in his bone. 'I went to lift my leg up, and as I did so the upper part of the leg came away without the bottom half. The lower leg stayed completely flat on the bed. I could see my skin stretching.' Paul knew he would have to wait for his flatmate to come home before calling for an ambulance, as he would not be able to come to the door to let them in. When an ambulance crew finally arrived in the morning, Paul announced he had broken his leg. 'I very much doubt you've broken your leg,' the paramedic replied. 'You'd be in a lot of pain right now.' Despite Paul's explanation, she refused to believe him. 'I thought, *I'm not going to argue with her*, so I pulled back the covers and lifted up my leg. She could see the stretching of the skin, and turned pale: "Oh my God, you've broken your leg!" I said, "I told you!'"

The lack of understanding of Paul and Vicky's condition has been tempered by the generosity of strangers. Donations have come in from around the world by people captivated and horrified by the stories of children growing up without this most vital sensation; phone calls from mysterious donors in Saudi Arabia; local charities raising money – all to enable Bob and Christine to try to bring up their children as normally as possible.

Even when listening to Paul explain it to me, the impact of a world without pain is so alien that I still cannot quite understand it. Paul and his sister's experience of life is so utterly different, fundamentally so, from most of humankind's, that I am not surprised by Paul's inability to empathise with people in pain. 'Like trying to teach a blind man about colours,' Bob says at one point. Given the rarity of their condition, for much of their lives the only people who truly understood the brother and sister's world were each other. But with the advent of the internet, Paul has found someone else who understands him. Steven, based in Washington state in the north-west corner of the US, is Paul's 'brother from another mother'. Steven also has CIP, and Paul chats to him every week. Their lives are a mirror image of each other. 'What I've experienced is what he's experienced,' Paul says: a shared perception of reality as a world without pain. Similar injuries, similar stories from childhood – and similar tragedy. Paul has lost one sister, and, as we chat, Vicky, his other sister, is recovering in hospital, having undergone an amputation of a leg so damaged, so mangled, by repeated injury that it was rendered useless. For Steven, it was the suicide of his older brother, also living with CIP. Repeated fractures of his vertebrae caused him to develop compression of the spinal cord, slowly robbing him of the use of his legs.

The loss of a previously active outdoorsy lifestyle – his favourite activities were hunting and fishing – resulted in Steven's brother taking his own life. For Paul and Steven, identical physical scars and psychological wounds are the signature of a life without pain.

A life without discomfort, ache or agony is beyond the comprehension of most. However, even for a person with an intact neurological system, moments of absence of pain, even in the context of massive injury, can exist. 'There is a common belief that wounds are inevitably associated with pain, and, further, that the more extensive the wound, the worse the pain. Observation of freshly wounded men in the Combat Zone showed this generalization to be misleading,' began Lieutenant Colonel Henry K. Beecher in the opening of his paper entitled 'Pain in men wounded in battle', in 1946. Beecher had been an anaesthetist in the US Army during the Second World War, working in the field in the Mediterranean Theater of Operations. In his paper he described his recent experiences of the care of soldiers returning from the Venafro and Cassino fronts in Italy and from the battlegrounds of France. Detailing soldiers' descriptions of their pain from terrible and devastating injuries – penetrating injuries of the head, chest or abdomen, compound fractures of their limbs or extensive soft-tissue injury – he wrote about 215 patients. What he found was staggering: despite their bodies being shattered by bullets or ordnance, less than a quarter of soldiers reported severe pain, and three quarters of men did not ask for pain relief, even though they knew it was available. He wrote:

This is a puzzling thing and perhaps justifies a little speculation. It is to be remembered that these data were obtained entirely from wounded soldiers. A comparison with the results of civilian accidents would be of interest. While the family automobile in a crash can cause wounds that mimic many of the lesions of warfare it is not at all certain that the incidence of pain would be the same in the two groups. Pain is an experience subject to modification by many factors: wounds received during strenuous physical exercise, during the excitement of games, often go unnoticed. The same is true of wounds received during fighting, during anger. Strong emotion can block pain. That is a common experience. In this connection it is important to consider the position of the soldier: his wound suddenly releases him from an exceedingly dangerous environment, one filled with fatigue, discomfort, anxiety, fear and real danger of death, and gives him a ticket to the safety of hospital. His troubles are about over, or he thinks they are. He overcompensates and becomes euphoric . . . On the other hand, the civilian's accident marks the beginning of disaster for him. It is impossible to say whether this produces an increased awareness of his pain, increased suffering; possibly it does.

As a curious footnote, he also comments:

A badly injured patient who says he is having no wound pain will protest as vigorously as a normal individual at an inept venipuncture [the act of drawing blood]. It seems unlikely that the freedom from pain of these men is to be explained on the basis of any general decrease in pain sensitivity.

Many of us will ourselves have experienced that the degree of pain is influenced not just by the nature of the injury, but by other factors too. The pain of an injury is less noticeable when we are distracted; an aching joint is more intrusive when we are tired or anxious; the limp only surfaces after coming off the playing field; the scrape to a knee or elbow eases with the rubbing of the skin around it. These are based on our previous experiences and memories of pain, our expectations of pain to come. Our state of mind influences our experience of pain. At face value, it's an interesting phenomenon but not too noteworthy. Yet it tells us a huge amount about our perception of pain; in fact, not just pain, not even just sensation, but every one of our senses. It shows us that the act of sensing our environment is not simply the act of passively absorbing information, not just the flow of impulses from the external world into the internal. It illustrates that information also flows in the opposite direction, that the interior has an important influence on the transmission of data from the outside in. This crucial aspect of our nervous system is something we will return to several times over the course of the following pages.

It is these neural pathways that also underlie the placebo response – the improvement of pain with a sugar pill. When it comes to analgesia, approximately one third of the population will experience a significant effect on pain from a placebo drug. But the placebo response is not simply 'mind over matter', a psychological effect. If you take people who have had a response to a placebo and, without their knowledge, give them naloxone, a drug that reverses the effects of opioids like heroin and morphine, they will no longer have any benefit from the placebo. This clearly demonstrates that somehow the placebo is having a chemical effect, mediated by the body's own naturally produced opioids – chemicals similar to morphine – secreted within the nervous system. Block the effects of those opioids and you block the effects of the placebo. Indeed, neuroscientists can even take brain images of the placebo response. Scans can directly visualise the activation of specific receptors for opioids in particular parts of the brain – in that network of areas known to be involved in a wide array of functions such as cognition, emotion, motivation and, crucially, pain, and termed by some as 'the pain matrix'.

So, the expectation that pain will be relieved by a magic pill actually does lessen the experience of pain. But the inverse is equally true; the expectation of pain enhances it. I think of the countless times I have prepared someone for a lumbar puncture, the procedure through which a couple of tablespoons of spinal fluid are extracted. The patient lies on their side, facing away from me, curled up in a ball to open the spaces between the vertebrae of the lower spine. After a little local anaesthetic, I insert a long but fine spinal needle, finding the fluid bathing the nerve roots within the spinal canal. But for the more anxious or needle-phobic patients, the pain starts even before the local anaesthetic has been injected. The expectation of a needle can sometimes result in them jolting with pain or letting out a cry as I place the cold iodine swab on their lower back to sterilise the skin before any sharp object goes anywhere near them. This can be lessened by long conversations about the procedure itself, explaining as I go along, relaxing the person and lowering their anticipation of the intensity of pain – but this doesn't always work. Essentially, this is, in contrast to a placebo, a nocebo response.

The nocebo response, too, has been studied in a scientific context. In one study, sixty healthy individuals were subjected to pain by applying a tourniquet around their arm and asking them to exercise the arm starved of blood. This process causes a build-up of lactate in the muscles, inducing severe muscle pain that normally becomes totally unbearable after 13–14 minutes.

The volunteers were divided into groups, receiving nothing at all, ketorolac – a non-steroidal painkiller similar to, but more potent than, ibuprofen – or normal saline, or a combination of both. But the groups were told different things regarding the normal saline. Some were told it was ketorolac, inducing the expectation of a pain-reducing effect, while others were told it was a substance that increased pain. The subjects given nothing at all tolerated the pain test for some 13–14 minutes, as expected. When participants were given ketorolac, they managed to endure it for significantly longer, on average 22–25 minutes. But the expectation of what they were being given also clearly had an influence. Told that the normal saline they were getting was a painkiller, subjects managed 16–18 minutes, longer than normal. But those told that the saline was a drug that heightened pain managed, in some cases, as little as nine minutes. And further studies have demonstrated changes in activity in the areas of the brain implicated in the 'pain matrix' induced by the nocebo response, as they have for the placebo response.

It is obvious, therefore, that when it comes to pain, the positive or negative expectation of pain influences pain itself. But it is also clear that this is not purely a 'psychological' phenomenon and that our anticipation of discomfort directly influences brain activity and chemistry. Yet how do these changes in the brain itself alter our perception? Is it purely that the brain has altered its own activity, so that the areas of the brain that 'feel' pain are inhibited or excited by these other factors? Is it just the effects of endogenous opioids – those produced by the body – or other chemicals like cannabinoids (other endogenous chemicals similar to the active components of cannabis) dulling brain activity? In recent years, we have learned that these soothing effects are not happening just at the top of the nervous system, but that actually the pain-modulating effects are also exerted, at least in part, much lower down in our bodies. A series of regions deep in the brainstem, far away from the brain's outer mantle, the sensory cortex – where we 'feel' pain – are perhaps responsible for many of the effects of morphine, and the opioid chemicals produced within the brain itself. One region in particular, the periaqueductal grey (PAG), located roughly halfway between the cerebral cortex and the spinal cord, is of fundamental importance to this process. Injecting morphine into this area, or stimulating it electrically, causes a powerful painkilling effect. But evidence also points to the underlying effects on pain being brought about within the spinal cord itself. From the PAG and associated regions, nerve fibres project down the spinal cord, modulating pain signals at the point of their entry into the spinal cord, much lower down. These projections directly influence the flow of 'nociceptive' (pain sensation) impulses into the central nervous system, essentially being able to dial up or dial down pain far away from the brain. Damage these projections and this ability is lost, with electrical stimulation of the PAG, or morphine trickled into it, rendered useless.

These circuits are fundamental to the human experience of pain and the response to pain-numbing salves. But it may also be that these mechanisms of control, from the brain to the spinal cord, are responsible for many of the chronic pain syndromes that we see in clinical practice – conditions like irritable bowel syndrome; pain resulting from a previous injury, long ago healed; pain – severe, unrelenting, life-changing – persisting long after its original cause has disappeared, or was never there in the first place; pain caused by plasticity, changes in the nervous system and its organisation, causing the amplification of pain or even the interpretation of non-painful sensory stimuli as pain, much like the patient awaiting their lumbar puncture, who flinches at the first touch of the iodine swab.

We have known for many years that certain circuitry in the spinal cord influences the perception of pain. I recall, as early as my first year of medical school, being taught of the physiological basis of rubbing a painful knee or scratching around a cut to relieve pain; learning that, when stimulated, fibres conducting non-painful sensations dampen down those bundles of nerve fibres conducting pain. This is the so-called gate-control theory; essentially, stimuli of light-touch, scratch or temperature shut the gate on the flow of pain impulses, limiting the number of those signals reaching our brain. But what we know now is that the brain also directly influences this gate, opening or closing it depending on expectation, memory, anxiety and a range of other factors.

And what of Beecher's mangled soldiers, surprisingly untouched by the severe pain normally caused by such horrific injuries? Experimental studies have shown, in animals at least, that acute intense stress has an analgesic effect. But this effect is nullified by both damage to the descending fibres, radiating from the brainstem down to the spinal cord, and the injection of the opioid-blocking drug naloxone. So, it seems that the painkilling effect of stress is mediated by opioids and those descending fibres. (In truth, it is likely that other systems are also involved, such as endogenous cannabinoids.) It is these mechanisms that likely underlie Beecher's reports of his soldiers and their lack of pain, their euphoria even – as well as accounts of people continuing to play in a sporting match despite a broken bone or twisted ankle, the pain only becoming apparent when the excitement has passed. From an evolutionary perspective, these are life-saving processes within the nervous system that enable us to run away or fight even when injured, but also enable us to experience pain when the peril has gone.

I ask Paul what he would do if offered a cure, a way to feel pain. His response surprises me. 'A lot of people have said to me, in the past, that it must be brilliant not to feel pain, not to worry about injuring yourself because you're not going to feel the pain of it. And my answer is always that if I could turn back the clock and be normal – feel pain – I would. But if I was given a cure now, I wouldn't necessarily take it, because, for me, the damage is done. I don't think I could cope with the damage I have already sustained.' With every joint and bone in his body riddled with damage, Paul is already crippled by limitations to his movements and his walking, but at least the absence of pain makes it more bearable.

If there is any chink of light in the darkness of the family's world without pain, it is the success story in our understanding of how a genetic change results in this condition; how a tiny change in a gene can switch pain off, how fundamental the role is of this sodium channel, Nav1.7, in the transmission of pain impulses. This knowledge opens the door to new therapies for those individuals who feel excessive pain. It is a measure of Paul when he says, 'From my point of view, if I can use what I have to help people that feel too much pain, I'm all for it. That is one good thing that can come out of the very negative thing I've grown up with.'

As the end of my visit approaches, Christine sits down at the table. Throughout our conversation, I have felt her reluctance to talk, to open up, and have interpreted it as suspicion. But while she sits there next to me, I can see her emotion, an overwhelming sadness, as she listens to Paul and Bob. And at one point, after we have been chatting for an hour, in a soft voice so quiet that I almost miss it, she whispers, 'I just feel so guilty. I feel so terribly guilty. That I passed this on to my three children. That it's my fault.'

*

While pain is the bully boy of touch, undeniable and unignor-able, this sense has other modalities, too. A gentle breeze shifting the hairs on the back or your neck, the feel of a cold glass of beer in your hand, the vibration of your mobile phone in your pocket – less demanding of your attention than the searing agony of a broken bone or the soreness of a paper cut, but nev-ertheless clear perceptions of sensations on your skin. Yet there are some 'sensations' that do not even enter our awareness, that float through our bodies constantly but are never recognised – at least, not until they are gone, when they are only notable through their absence. And when these sensations are lost, the consequences are life-changing.

Whenever Rahel comes into my clinic, her size is a barom-eter, an indication of the weather outside. On the warmest summer day, she wears a couple of thick, brightly coloured jumpers. In winter, she is enveloped by several layers of woolly clothes, a thick overcoat and huge, oversized mittens. Her almost black eyes peer out of the space between her purple ski hat and her high collar. But beneath the volumi-nous fabric, she is skin and bone, her cheeks sunken, a fragile sparrow in a rainbow palette of coverings. It is only when I first see her at home, some six years after we've first met, that I appreciate quite how little of her there is. She takes up almost no space in the room, and looks like she would snap in a gentle breeze. In all these years, it is the first time I've seen her bare head, normally hidden under a turban or a woolly hat; braids of grey tightly follow the contours of her head. But as I sit in her living room, I spot evidence of the woman she was before I met her: a picture of a young woman in a white wedding dress, gazing directly into the camera, brimming with happiness; a photo of her with her husband, another of him in military uniform, a certificate of military service from a Middle Eastern royal; an Arabic coffee urn, a spider embedded in plastic, bigger than my hand – the stuff of nightmares. Documentation of a life well-travelled, journeys to exotic lands.

'I came to England when I was thirteen', she tells me. 'My mother was from Eritrea, my father from Ethiopia. They came here for my education, but two years later the Trouble started and we couldn't go back.' In 1974, the Emperor of Ethiopia, Haile Selassie, was overthrown in a coup by the Marxist-Leninist Derg military junta, and return was impossible. 'I met Roger at the age of eighteen, and we got married when I was nineteen. We have been married for forty-one years,' she laughs ruefully, a reminder of the passage of time. Her marriage took her back abroad over the years, her husband's service in the British Army moving them all over the world. Roger walks quietly in and out of the living room, trying not to disturb us, but even in his few steps I see a regimental bearing, and there is a clear imprint of a life in the military in his voice.

I recall that when I first met Rahel six years ago, I knew her by a different name. 'When I first started coming to the hospital, they called me Rachel,' she explains. This error has only been corrected in the past year, and I have noted the change on her medical records. She has finally reclaimed her own name, in the last few months of her life. I tell her that my grandparents and my mother were also renamed by bureaucratic clerks, my grandparents dying with names almost entirely unrelated to those that they were born to – testament to lives disrupted by conflict, war and relocation. Rahel makes a wry upturn of the mouth in recognition.

Even when I first met Rachel/Rahel, she and I knew she was dying. Then in her mid-fifties, she had been diagnosed with lung cancer, a type known as small-cell – malignant and malevolent, a tumour known to spread quickly, to bone and brain. She has been through chemotherapy and has been offered radiotherapy to the brain and spinal cord – a toxic blast of radiation to the nervous system – in the hope of killing any invisible cells that might have already sought refuge there. The cancer has not fully responded to the poisonous drugs infused into her veins, and she has declined radiation, feeling weak and exhausted from the gruelling treatment she has received already and not eager to put herself through it just as a preventative measure. Her oncologist has discharged her to palliative care, for the monitoring and treatment of symptoms as and when they arise. 'They told me I would likely live a couple of years,' she recounts now. 'And that was eighteen months ago.' Like a Buddhist monk, she smiles as she says this, as if she has totally accepted her fate.

But the reason that she first arrives in my clinic is somewhat different. She is unable to walk. Her balance has gone and each tentative step is a journey into the unknown, uncertain if she will find her footing or end up in a heap on the floor. I now no longer recall whether she was in a wheelchair at that first appointment or using a walking frame, as over the intervening years I have seen her enter my clinic room in a variety of ways – with one stick, or two, a rolling frame or, occasionally, under her own steam, unassisted by any walking aids at all.

As I read through her notes and hear her story, a number of explanations race through my mind. We tend to consider the act of walking a simple one, an unconscious act that can be done while talking, listening, thinking, eating. But this automatic behaviour belies years of learning, of development of our nervous systems and musculature. We do not stand on our first day of life and stride across the room. Our first, tentative steps only happen some twelve months after birth, and our first few years of life are marred by bumps and scrapes as we toddle and fall. And the systems required to steady ourselves, to walk with a confident gait, are countless. We obviously need strong legs and core muscles, with functioning joints and straight limbs. But it is the control of these legs that is key. The nervous system needs to be able to regulate the delivery of power to these muscles in a highly coordinated way. Those systems running from the brain's motor cortex all the way to the muscles need to be intact, from the brain itself down the spinal cord and to the motor neurones – the peripheral nerve fibres delivering signals to the muscles themselves. We also need to be able to coordinate the leg movements. Clearly a baby has the strength to be able to hold a leg extended to support its own bodyweight, but without the ability to move both legs at the same time, walking across the room may as well be walking on the Moon. The brain's ability to process gravity is also key. If you don't know which way is up and where you are in space, walking in a straight line is a hopeless task. We will all have had experience as children of spinning furiously around on a merry-go-round in the playground, then suddenly stopping to get off; the sensation of the world spinning, despite your feet being firmly planted on the ground, results in much hilarity as you are flung to the floor or weave your way over to the swings. Reasonable vision also helps; being able to see where you are walking – undulations in the ground, the nature of the surface – all allow for a firm foothold.

As we continue to talk about her symptoms, Rahel tells me that since this first started, she has also felt some numbness in her hands and feet. Not a total loss of sensation, more a wooliness of touch. When she holds objects – a cup, a door handle – the certainty of her grip is not quite there. And when I begin to examine her, there are more clues as to what might be causing her walking difficulties. I start by looking for indications of problems with her balance or coordination. For Rahel, there are multiple possibilities why these might be damaged. Most pressingly, the possible spread of her cancer to the brainstem may be causing the signals from her inner ears to be disrupted, confusing her brain as to whether she is still or spinning. A metastasis in the cerebellum – the 'little brain' above the nape of the neck, responsible for coordinating our movements – may also cause this. And her chemotherapy for lung cancer may also have damaged the cerebellum. This part of the brain is often vulnerable to the effects of chemicals. Think of a Friday night in any city centre, full of stumbling, slurring pub-goers, alcohol directly preventing the cerebellum's normal function.

But, assessing Rahel, I find no features of increased pressure in her head – a warning of cancer inside her skull – nor are there clear signs of problems with her cerebellum. However, as the examination continues, a potential explanation is forthcoming. I ask her to hold her hands out in front of her, fingers outstretched. Almost rock-steady, the fingers remain firmly still. But then I ask her to close her eyes, and as she does so her fingers, previously still, begin to wriggle uncontrollably and her arms begin to drift up and down. When I ask her to stand in the middle of the room, with me carefully placing myself behind her, she rocks back and forth a little but remains upright. Once again I ask her to close her eyes, and immediately she veers back and forth with violence before collapsing into my arms. When I examine her limbs, apart from some very mild loss of light-touch, there is little to find – until I assess her ability to detect movement, that is. With Rahel lying on the couch, eyes closed, I grip the tips of her fingers and toes, moving them up and down. But when I ask her to tell me which direction I have moved her digits, she has absolutely no clue. And as I work my way up, testing her ankles, wrists, knees and elbows, there is nothing, no ability to detect movement. Even at the shoulders and hips, she can only perceive very large movements. This aspect of her sensation, known as proprioception, has almost entirely vanished. And it is this absence that is as telling as what is present, like Sherlock Holmes' 'curious incident of the dog in the night-time'.

Without sensation, it is difficult to move. We think of such functions of our bodies as entirely separate, independent of each other, but in reality they are intimately linked. Consider the act of picking up a glass. The muscles of our shoulder, arm, wrist and hand act in unison to move our limb into the correct position as we curl our fingers around the glass and lift it to our lips. But how do we avoid breaking the glass by exerting too much pressure, or letting it slip from our hand by not grasping it tightly enough? We of course feel the pressure-sensation on the pulps of our fingertips and the palms of our hands, and years of practice make us less clumsy, knowing precisely how hard to hold things. But it is not only this aspect of sensation that is of vital importance for our movement. If you hold that glass and close your eyes, you can still lift it to your lips and take a drink; your body knows where your limbs are in relation to each other and in relation to your external world, regardless of whether you can see your body parts. Otherwise, how could you scratch an itch behind your ear, walk in the dark or touch-type without looking?

This sensory information is so important that without it we are rendered almost useless. Entire tracts of nerve fibres, in our peripheral nerves and spinal cords, are dedicated to it – sensors all devoted to the perception of tiny movements or changes in position. Receptors in the joints or in the skin overlying the joints provide important information about how flexed or extended each joint is. But perhaps most important is a small structure known as the muscle spindle receptor, twisted around specialised muscle fibres. This tiny structure is exquisitely sensitive to minute changes in muscle length, and provides the all-important feedback caused by active or passive lengthening or stretching of muscles depending on body position. In fact, the muscle spindle receptor is responsible for the most iconic test that we neurologists routinely do.

If there is one thing that patients expect of me, whether it is necessary or not, it is the wielding of a tendon hammer and the knee-tapping to check their reflexes. As they walk into clinic, they see that round, chrome-and-rubber head and long, white, plastic handle on my desk, and I sometimes get the feeling that people feel short-changed if I do not attack their knees with it during their visit. What we are actually doing with it is stimulating the muscle spindle receptors. By tapping a tendon, we very briefly and rapidly stretch the muscle, stimulating the feeling that the limb is being passively moved. As a reflex response, the muscles instantly contract a little to maintain body position, which results in a jerk. In this way, we test the circuitry underlying this reflex – the sensory signals conducting muscle-stretch information to the spinal cord, and the motor signals back down to the muscle. But there are other tricks that can be played. The spindle receptor is also powerfully stimulated by vibration. Apply vibration to a muscle and this will result in the perception that the muscle is being stretched. As with the tendon hammer, application of vibration results in the illusion that your limb is being moved.

The simple act of movement needs constant adjustment, with ongoing corrections, depending on the reality of the current status of our body. Take the lifting of that glass of water to your lips. If you do it over and over, with the glass and your body in the same position, in theory your muscles will learn to exert the precise series of forces on your arm to achieve this success-fully. But then put on a watch or fill the glass a few millilitres more, and those same forces may not achieve the same result in the presence of tiny differences in mass. In the absence of knowledge of where your arm is in space, small changes in your world would result in wildly different outcomes, like splashing the water in your eyes rather than taking a sip.

I have seen cases like Rahel's before. I recall a few years ago seeing a young man, in his mid-twenties, working in the media. Rob (not his real name) was a man about town, enjoying the rich social life that London has to offer; confident, cheeky and surprisingly blasé about the fact that over the preceding week he had gone from partying throughout the whole weekend to being unable to walk in a straight line. Examining him, I found the same features as seen in Rahel, a complete inability to know where his limbs were in space, to such an extent that he could barely stand without support. He was otherwise entirely fit and well, without any obvious cause for his condition. I interrogated him about any recreational drug use and he denied taking any. When I asked him if he had been inhaling laughing gas (nitrous oxide), he looked slightly astonished. After a brief pause, he slightly sheepishly told me, 'Yes, I've used it a bit.' Irritated by his obtuse response, I pursued this line of questioning and asked him what 'a bit' meant. 'Well, about thirty balloons on Saturday, and forty on Sunday.' And as the story unravelled, he admitted to regular and heavy use of nitrous oxide.

As I cycle back home from the hospital, I pass through the Vauxhall area of London. This neighbourhood is rather unu-sual, even for London. At one end, at 85 Albert Embankment, is the hulking MI6 building, scowling over the Thames. Its architect was inspired by Aztec and Mayan temples, as well as 1930s industrial modernist styles, and there is also something reminiscent of Nazi-era buildings about it. With its roof covered in satellite dishes and antennae, it casts a somewhat dark shadow. At the other end is the new US embassy, another behemoth, a block-like fortress surrounded by a pseudo-moat. When Donald Trump heard of the plan to move the US embassy from the genteel environment of Grosvenor Square, in the heart of Mayfair, to this less salubrious part of London, he was disgusted. It was going from the most expensive area on the Monopoly board to one that does not even figure on it. Trump criticised the move, calling it a 'Bush–Obama special' and describing the location as 'lousy' and 'horrible' – for, sandwiched between the embassy and the MI5 building is Vauxhall Cross, the site of a large train station. On one corner sits the Vauxhall Tavern, a landmark in London's gay scene, famed for its drag shows since the 1970s. From about 7 p.m. onwards, a queue of people in drag and other audience members snakes around the pub, waiting to enter. In the arches under the tracks are located a series of clubs, bars and saunas, and from the hours of dusk, sometimes until nine or ten in the morning, the streets are awash with inebriated and chemically altered partygoers and pleasure-seekers. I suspect they are as fond of Trump as he is of them.

Passing through early in the morning, the road under my bicycle tyres often tinkles as I cycle; scattered across the ground, like the remnants of a glitter party or a sprinkling of silver snow, are countless small silver canisters, an inch or two in length, flattened by the passing traffic. These canisters will each have contained eight grams of nitrous oxide; they are made for catering companies, where they are used to make whipped cream. But for the partygoers of Vauxhall their purpose is somewhat different. It is used to inflate party balloons, which are then sold and inhaled for a few seconds of a high. Inhalation of nitrous oxide induces a brief period of euphoria, and occasionally mild hallucinations. It is also known as 'hippy crack' or, in the US, 'Whip-Its'.

But the regular use of nitrous oxide can have some unintended consequences. In addition to its desired effects on the brain, nitrous oxide has some chemical effects elsewhere. It essentially deactivates vitamin B12, a nutrient fundamental to the health of our nervous system and the generation of red blood cells. Essentially, regular and heavy nitrous oxide usage makes people profoundly vitamin B12-deficient.

There are other causes of B12 deficiency, most notably pernicious anaemia, a diagnosis that sounds straight out of the pages of a Dickens novel. Pernicious anaemia is different from normal anaemia, and is not due to iron deficiency; rather, it is caused by one's own immune system attacking the lining of the stomach, preventing the production of a chemical vital for the absorption of B12. No matter how much vitamin B12 you consume, it simply does not enter into the bloodstream, leading to a deficiency of usable B12 – 'starvation in the midst of plenty'. This results in severe anaemia unless recognised and treated while still in its early stages. Pernicious – meaning harmful, in a gradual or subtle way – refers to the slow, unnoticeable progression of the condition, until the anaemia is so severe that you are left weak and exhausted, with a racing heart and shortness of breath. But in this context, pernicious also means deadly, an inevitable endpoint in the era before effective treatment.

And it's not just anaemia that results from the lack of vitamin B12. There are neurological complications too. Patients will experience tingling in the hands and feet, as the nerves transmitting sensory information begin to malfunction. As the damage progresses, sufferers will gradually develop weak legs and increasing numbness. But the nature of that numbness is unusual. While pain and temperature-sensation are usually preserved, it is light-touch, the sensation of vibration and, in particular, a sense of joint-position, that are affected. There seems to be a disconnect between the various sensory modalities, with some sensations entirely unaffected and others obliterated. The reason for this is not immediately clear – that is, until you look closer. Historical specimens of people dying with this condition exhibit the hallmark of vitamin B12 deficiency. While the nerves are indeed affected, the spinal cord is as well. Looking at sections of the spinal cord under the microscope reveals changes there, but it is the location of these changes that explains this dissociated sensory loss. Various areas of the spinal cord show swelling, degeneration and gradual scarring, particularly in the part of the spinal cord known as the dorsal columns, the aspect of the spinal cord closest to the skin of the back, and comprising thick bundles of fibres transmitting impulses encoding vibration sense and joint-position sense from the periphery to the brain.

In the modern age of medicine, rather than just waiting for people to die of pernicious anaemia, we can scan them. MRI scans show a similar abnormality to historical examples, with changes visible in the dorsal columns – parallel tramlines of different sensory modalities existing in the spinal cord. Indeed sometimes a scan is how the diagnosis of B12 deficiency is made. So while the dorsal columns transmit certain sensations, others, like temperature and pain, are conveyed elsewhere, in the spinothalamic tracts, which are in the part of the spinal cord that sits much closer to our front than our back.

In fact, this organisational detail of our nervous system can generate some surprising results, and ones that are useful to neurologists, giving us important clues as to where the problem lies.

For instance, most people will be familiar with the idea that the left side of our brain is responsible for movement and sensation on the right side of our body; the right brain corresponding to our left side – that somewhere within our nervous systems there is a crossing-over of signalling from one side to another. But for different parts of the nervous system the site of the cross-over varies. When it comes to sensation, the tracts that conduct the sense of vibration and of joint-position switch sides at the top of the spinal cord, as it enters the brain. In contrast, the pathways conducting pain and temperature cross over at the point at which they enter the spinal cord, much lower down.

You would be forgiven for thinking these are irrelevant details, but they are important if you damage your spinal cord. If one side of the spinal cord is injured – by inflammation, compression or a foreign body like a bullet or a knife wound – you may end up with a very strange set of symptoms. Bizarrely, you may experience the loss of pain- and temperature-sensation in one leg, while in the other leg you lose sensation of vibration and joint-position – dissociated sensory loss, affecting both legs but in different ways. Imagine a knife stabbed through your back, cutting through the whole of the left side of your spinal cord at the level of your chest. At the level of the cut, the dorsal columns are carrying information about light-touch and joint-position from your left leg that is yet to cross over to the other side. But the spinothalamic tracts have crossed over below the level of the cut and are conveying signals of pain and temperature from the right leg. This condition, known as Brown-Séquard syndrome, is beloved by medical students due to its striking nature and equally cherished by those of us trying to educate those medical students. It provides a very concrete example of the highly organised nature of the nervous system and an important aide-memoire to neuroanatomy.

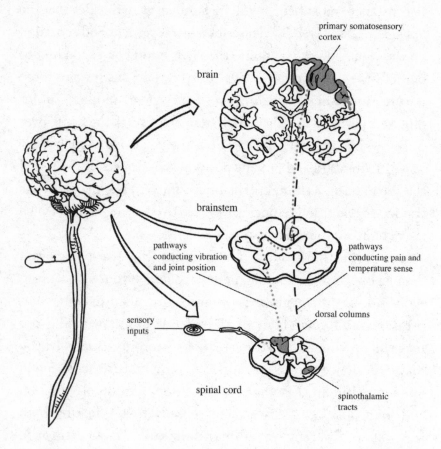

Figure 1. Sensory pathways to the brain. These pathways differ, depending on the sensory modality. Pain and temperature sensation is conducted by the spinothalamic pathways (dashed line) and crosses over to the opposite side of the body as soon as these impulses enter the spinal cord. In contrast, vibration and joint-position sense/ proprioception (dotted line) remain on the same side of the body in the dorsal columns, until they pass into the brainstem. Ultimately, all sensory data passes to the primary somatosensory cortex on the opposite side, but due to these two pathways damage to one side of the spinal cord can result in differential impairment to the different types of sensation, as in Brown-Séquard syndrome.

In Rob's case, however, there was no evidence of any impact to his temperature- or pain sensation. His inability to walk was almost entirely down to a lack of positional information; a scan showed the characteristic appearances of vitamin B12 deficiency, known as subacute combined degeneration of the cord. In fact, even before the scan and the results of blood tests confirming the lack of B12, I had given him daily vitamin B12 injections, and over the next few weeks his walking normalised. I discharged him a few days later, with strict instructions to use nitrous oxide only for whipping cream. Whether or not he took my advice, I will never know. After turning up in my clinic a couple of times, he subsequently disappeared into the sprawling mass of the city, nonchalant and seemingly without embarrassment, never attending any further appointments.

Rahel, of course, is entirely unfamiliar with the little silver canisters of laughing gas scattering the streets of Vauxhall. But it is clear that something similar is going on within her nervous system. The ability to detect movements in her limbs, to sense, either consciously or unconsciously, the relationship between the positions of her toes, feet, knees and hips – so crucial to the act of standing or walking – has been cryptically and critically impaired. She is left barely able to stand without assistance, and taking a couple of steps is almost beyond her abilities. But ultimately, the dark shadow, some 2cm in diameter, in the apex of her left lung, of small-cell cancer, looms over her, and over my diagnostic thinking.

Over the next couple of weeks, I subject Rahel to a battery of tests – blood examinations, scans, a lumbar puncture to examine her spinal fluid, electrical tests of her nerves. These nerve conduction studies show that there is an absence of sensory transmission through the nerves of her arms, and to some degree her legs too. The MRI scan shows no obvious spread of the cancer to her brain or spinal cord. But, like Rob and his nitrous-oxide-induced imbalance, when I look at the black-and-white images on my computer screen, there are changes apparent in the spinal cord, from the top of the neck to the middle of her chest level, tracking along the dorsal columns – the tracts mediating vibration and joint-position sense. Something has damaged her spinal cord as well as her nerves. But for Rahel, vitamin B_{12} is not the culprit – all her blood tests are normal and there is no evidence of inflammation or infection in her spinal fluid.

A few days after the tests have been completed, I see Rahel again in my clinic. She is pushed into the room on a wheelchair by a supportive friend. She smiles in greeting but she looks fragile and weak, still swathed in layers of wool and cotton – cardigan over jumper over sweatshirt over shirt. I have already spoken to her oncologist to learn what her prognosis is. She is for palliative care only. She has perhaps a few months, a year, maybe even two at a stretch.

As we chat, it is obvious that if anything she has got worse, now unable to stand at all. We discuss her results, and while I think I know what might be happening, I have no definitive evidence. So, to delay the moment when I need to admit I am not sure, I go through the results one by one. I list the normal results, and to each pronouncement she nods silently. I then go on to describe the electrical study report, and the findings from the scans. I tell her there is no evidence of spread of the cancer, to which she briefly smiles. I pause for a moment, before going on, as I know that what I say next is somewhat speculative, based on the overall clinical picture rather than any specific results.

Despite no evidence of the cancer progressing, I tell Rahel that I think the cancer is indeed to blame, albeit in a somewhat unexpected way. Cancer causes damage not just through invasion, through infiltration of normal tissue, through disrupting function; in some cases, cancer is the orchestrator of other modes of attack. Certain types of cancer, and particularly small-cell lung cancer, can trick the body into damaging itself. While cancerous cells come from our own body, and are therefore difficult for our immune systems to recognise as foreign invaders, sometimes the immune system does identify the abnormal cells as such, and this generates an immune response, an effort to rid the body of the disease creeping through our tissues and bloodstream. But certain aspects of the cancerous cells can look very similar to other cells in our body, causing the immune response to also target these bystanders in the battle between the cancer and our own mechanisms of protection. This results in damage triggered by the cancer but carried out by our own body – a sort of cellular suicide.

And as I tell Rahel all this, her eyes are fixed on my face, listening intently. Despite the failure to identify a specific antibody in her that is associated with this type of syndrome, I say that, nonetheless, this is what I think she has: an immune attack on her spinal cord and peripheral nerves, triggered by her cancer, causing this failure to conduct impulses for joint-position sense – proprioception – from her limbs to her brain. In normal circumstances, the treatment would be to rid the body of the cancer, but this has been tried to no avail. The only other option is to suppress her immune system, to dampen down the damage her own body is doing to itself. Theoretically, however, her immune system may actually be helping keep her cancer in check, and slowing it down may risk accelerating the growth of her tumour. I must have painted a rather bleak picture, and I recall proposing this treatment option couched in somewhat pessimistic terms. At one point, a few years later, she tells me, 'I remember that when you suggested trying something, I said, "We've got nothing else to lose. Please, anything that I can have to help me would be very nice.'"

We agree to try a specific treatment called immunoglobulin. In this treatment, a preparation of antibodies collected from thousands of normal individuals is infused directly into a vein every few weeks. The precise mechanism of this action is uncertain, but flooding the system with normal antibodies seems to dilute the unwanted immune response driving this condition. And so, with low expectations, on my part at least, we agree to go down this road.

I see Rahel four weeks after the infusion, and I am somewhat taken aback to see her walk into the clinic room unassisted, a look of pride on her face as she clutches a walking stick in one hand. She is effervescent, glowing with a sense of achievement. She tells me that she has gone from being virtually unable to stand, to walking for thirty minutes at a time, even braving a walk to the local shops. Her garden, neglected for the past few months, has now regained its order through her recovery. She tells me some time later, 'I don't know if you believed me. I thought, *He thinks I'm telling fibs. I don't know whether he believes this is working on me!*' But as I examine her, it is obvious that her coordination, the jerkiness of her hands and her ability to stand with eyes closed all are better, if not totally normal. My pessimism was clearly misplaced.

Over the next few months, we settle into a pattern. As the infusion wears off, her walking and co-ordination deteriorate, a signal that another is required. Every eight weeks, she comes into hospital for a repeat infusion, and within a week or so is back on form. Each time I am due to see her in my clinic and I see her name on the list for the day, there is trepidation that she will not appear, the treatment having accelerated her cancer. But each time, as I go into the waiting room, I am greeted with

'Hello, Dr Leschziner!', though she struggles to pronounce my name – a feature of my name rather than her disease. And the months slowly turn into years, with our familiar dance of chat, assessment and approval of the next few sessions of treatment. Six years after we first meet, she is still walking into my clinic, still functioning, still independent.

The last time we meet, it is at her house. I am greeted at the door by her husband, whom she has talked about but whom I have never met. I am ushered into the living room, where she sits, and I can immediately see that there is a new edge to her frailty, her cheeks more hollowed, her eyes more sunken. She has known that she is dying for the past six years, but the news that she has been living in expectation of has finally arrived. Investigations for abdominal pain have led to the discovery that her cancer has spread, and once again the decision has been made to give palliative care. 'I've decided not to go for any more treatment,' she says. 'It's just too painful and I'm very weak; I can't even go to investigate this lump. And as it's not giving me any pain, I said I'd rather live with it for as long as I can.'

As we chat, and I look around the room, I see a life well-lived, of adventure, of love, of family. I realise that although we have spent time together, most of our interactions have been focused on the practical. There are other things that have gone unsaid, stories untold. Despite her current situation, she is accepting and philosophical. She has had many years to consider her mortality, and has outlived everyone's predictions, including her own. Discussing her decision to abstain from further treatment for her cancer, she says, 'I'm quite happy with that. All in all, I've had a good innings with all that I'm carrying. I've had a good life. I've travelled all over the world, I've had an exciting life, thanks to my husband. And I've got wonderful family. I'm taken care of, you know? I'm comfortable.'

As I depart and say my goodbyes, I am certain in my own mind that Rahel and I will not meet again. And a couple of months later I receive the expected news – that Rahel has passed away; her date of death a day after her date of birth. Sixty-one years and one day. 'A good innings', Rahel described it, but it was a short innings too, and for the last six years hampered by her nervous system, betrayed by her senses. Proprioception, a sensation that we are mostly unaware of, does not enter our minds day to day or even year to year – a stark contrast to the sensation of pain. Much like Rahel, quiet, unassuming, just getting on with it, it is only in its absence that its mark is felt.

There is something else we can learn from Rahel and Paul – that sensation is not simply the act of touch, the ability to feel hot or cold, dry or wet, soft-touch, pinprick or pressure; that it is not even one sense; that various aspects of our sensation, whether obvious or not, define our understanding of the rules that govern our world; that certain actions are hazardous to life and limb, detrimental to our survival, damaging us both physically and psychologically; that even the act of combatting gravity, by standing up, is an extremely complex activity and as much reliant on sensation as on muscle strength; and that we are all vulnerable to profound alterations to our perception of our world through tiny changes to our nervous system, an infinitesimally small disruption in our genetic code or a tiny glitch in our immune system, turning our lives upside down.

2

ZOMBIE FACES

'... just because you see something, it doesn't mean to say it's there. And if you don't see something it doesn't mean to say it's not there, it's only what your senses bring to your attention.'

Douglas Adams, *The Hitchhiker's Guide to the Galaxy*

True to form, and despite the glass roof of Piccadilly station, as I get off the train in Manchester, I can feel spots of rain somehow making their way onto my face. It is raining indoors, droplets splashing through the vents above my head. This is the city of my upbringing, where I spent my formative years. I feel disloyal writing it, but until I left Manchester to go to university, I genuinely thought that the sun almost never shone in the United Kingdom, that somehow Britain only had three seasons, spring showers dissolving seamlessly into the damp of autumn leaves. I am sure that my memory is unfair, but sitting on the lawns of Oxford in the summer sun, it seemed like a different country. And today the weather does nothing to dissuade me of my prejudices. It is filthy cold, with horizontal rain that no amount of waterproofing can resist; a life-sapping, penetrating damp chill that one rarely encounters in London.

I get on the tram and, as we pass through the city centre,

through the rain-streaked windows I can see the industrial buildings so familiar to me. Cavernous red-brick warehouses and mills, witnesses to the city's central role in the global cotton industry. As we pass over a canal, the water surface is rippled by raindrops. Damp patches spot the painted graffiti under bridges, water seeping through the walls. Fuelled by the cotton-spinning mills of Cheshire and south Lancashire, the city expanded rapidly in the nineteenth century to become one of the lights of the Industrial Revolution. Canals were dug to bring freight directly into the city, vast palaces of industrialisation were built, and Manchester became rich. This pedigree of technology persisted beyond its heyday – the city with the world's first intercity passenger station went on to become the site of the first splitting of the atom by Ernest Rutherford, the world's first digital computer and, as recently as 2004, the discovery of graphene. It was this technological base that brought my family here from Germany in 1980, my father's academic career drawn to the University of Manchester. The worker bee, adopted as a symbol of Manchester 150 years ago, adorns buildings, lamp posts and street bollards, denoting the work ethic, industry and hive of activity that Manchester represented. In the aftermath of the terrorist incident in 2017, at Ariana Grande's concert at the Manchester Arena, killing 23 and injuring 139 concert-goers, the bee symbol became a show of public unity and pride in the city, with large numbers of Mancunians having it tattooed on their skin as part of a fundraising campaign.

As I trundle through the city-centre streets on the tram, I can see a resurrected city. Coffee shops, restaurants, the previously pollution-stained buildings licked into shape, a few high-rise towers smattering the skyline. Even now, however, a mere stone's throw from the city's beating heart, there are patches of neglected wasteland, signs of a rejuvenation project still in progress.

In the 1980s, the era of my schooldays, Manchester had been in decline for quite some time, beset by social problems, poverty and crime. The city centre was very rough around the edges, and those massive brick buildings proudly proclaiming their glorious past were mere shells. I recall catching the train home, passing the old factories and warehouses lining the tracks: building after building, each several hundred metres long with multiple storeys, with literally every pane of glass broken through vandalism or neglect; like ex-heavyweight champions of the world, long past their prime, muscle gone to fat, faces scarred and battered, teeth all knocked out. But in the midst of all of this, Manchester was the location of a glorious revival of another sort. The grim environment, the dire economic climate, coupled with the strong Mancunian sense of fun, led to an escape in the form of music. Starting with The Smiths, New Order and Simply Red, by the late '80s and early '90s the bands that constituted the soundtrack of my teenage life were the Stone Roses, Happy Mondays, James, Inspiral Carpets, 808 State and, later, Oasis. Through this 'Madchester' cultural scene, the pride of Manchester was restored, this time through music rather than the smoking chimneys of the Industrial Revolution. The city was re-cast as the coolest place on earth, in our own minds at least. Boddingtons, 'the Cream of Manchester', was the beer of choice in pubs throughout the UK, or so it seemed to us at the time. I hear the opening bars of a Stone Roses song and I am immediately transported back to being sixteen and deeply uncool in the midst of the city.

I get off the tram in Chorlton, a couple of miles away from my old school. By the time I get to the house that is my destination, I am soaked through, and as I stand in the porch waiting for the door to open, water drips off my jacket, my trousers sodden. A woman a few years younger than me opens the door, smiles and ushers me into the dry warmth of the house. She introduces herself as Nina. I feel a little guilty as puddles of water form on the polished wooden floorboards. The house is dark as there are no lights on, but I can see it is beautifully decorated and immaculately tidy. Once dried off, jacket hanging up, I follow Nina into the modern kitchen and she suggests a cup of coffee, an offer I take up eagerly. While she potters around the kitchen, I have an opportunity to survey my surroundings. As I look around, I see I am inside a homage to that Madchester music scene. The walls are adorned with album covers, artworks, a portrait of the Stone Roses' lead singer, Ian Brown. In one corner, there is a colourful frame surrounding a painted maraca, instantly recognisable to someone of my vintage and origin. I comment upon it. 'Yes, that is one of Bez's,' Nina tells me, referring to the dancer, mascot and maraca player from the Happy Mondays, famed for his bizarre, wild-eyed gyrations, fuelled by huge amounts of mind-altering substances. Nina tells me, 'One of our friends is an artist who decorated lots of his maracas. It was a gift.'

But as I look around further, I notice signs of the reason why I am here. As Nina makes the coffee, she clips small devices onto the sides of the mugs, and as she pours from the kettle, they beep as the mugs are filled. Next to the kitchen counter, attached to the wall, is a series of hooks from which hang three white sticks. And this is the reason why none of the lights are on – for Nina is almost completely blind. Lights on or off, it makes very little difference.

As I sit opposite Nina, I can see that while her right eye is clear, her left is subtly different, orientated towards a slightly different direction. I soon learn that the clear right eye is not an eye at all. It is a prosthesis, a false eye. In all other regards, she looks totally unremarkable – slim, unlined face, clear skin, in a brightly coloured jumper and jeans. She speaks with the vowels of the north-west of England, an accent infrequently heard in London that warms my heart, like a smell reminiscent of childhood, familiar and comforting. And given the way she moves around the house, her lack of vision is imperceptible to an onlooker.

It wasn't always like this. When she was born, Nina's vision was normal. But at the age of two, she caught the flu. Although she recovered, the virus seems to have triggered inflammation within her eyeballs called chronic uveitis, slowly robbing her of her sight by scarring her corneas, the clear windows through which all light enters. I ask her what she remembers of her childhood. 'I don't remember much from when I was a child, especially when it comes to my sight; I remember it varied a lot over time. I went to a primary school, a special school purely for blind or visually impaired children. My memories are mainly of school and the great attention and help I got.' She remembers being taken out of her first year of secondary school for surgery, at the age of twelve. 'I had a corneal transplant and a new lens. They told me that the cataract was stuck to the eyeball, so they had to cut it out with the cornea.'

I ask her what her vision was like as a child, if she could recognise people's faces. 'I was very short-sighted,' she explains. 'People had to be close to me for me to recognise them. The vision fluctuated. When it was at its worst, it was all very foggy, like looking at a steamed mirror in the bathroom. And I had to use magnification for large print.' In her teens and twenties, however, as technology advanced, Nina went on to have a series of further operations. 'From that first surgery, when I was twelve, I went on to have five more corneal grafts. But the grafts kept on rejecting.' After each procedure, performed on one eye at any one time, her vision would improve – 'the fog, the steam had gone; my vision was a lot crisper and clearer' – but as the corneal grafts were rejected by her own body, the immune system kicking in and damaging the tissue of strangers, her vision would start to blur and fog up again. With each transplant came the promise of better sight, only to be followed by bitter disappointment again in the months afterwards. A rollercoaster of hope and despair.

Her last transplant, at the age of twenty-five, was different. She was referred to the world-famous Moorfields Eye Hospital, in central London. Just off the horrendously busy Old Street roundabout, the original Victorian façade hides a rabbit warren of clinics, operating theatres and testing facilities delivering the best treatment that ophthalmology can provide.

The surgeon Nina had been referred to had begun to use par-ticular drugs to suppress her immune system alongside these transplants; an effort to stave off rejection. She underwent a further corneal transplant on the right eye in 2008. 'It went really well,' Nina says. 'Straight after the graft, I was very wary – *Oh, it's just going to fail, like it has done every other time.* But afterwards, I had amazing vision. I could read a newspaper for the first time in my life without having to use a magnifier. It didn't only clear the vision; it enabled me to see smaller print as well. I was over the moon.' Despite side effects from the medication, the improvement in her vision was worth it. And this state of affairs continued.

'So, a year to the week from me having that surgery,' Nina chuckles nervously, then pauses, 'I had an accident at work. I was working in an office. There was a metal leaflet-dispenser stand in a box, stacked on top of a printer. The box fell, and one of the dispensers caught the corner of my eye.' She pauses again, but then continues to tell her story without a hint of emotion. I am shocked, not only by the turn of events but also her ability to tell me this without dissolving into tears. 'It burst my eyeball, and they couldn't save it at all. It was definitely a life-changing moment – 2009, January the twelfth. I remember the date.' The right eye, the recipient of this last corneal graft that had held so well, was now destroyed beyond hope of rescue by a hideous twist of fate, to be replaced by a prosthesis, an artificial eye. Nina continues: 'The surgeon said that he would look to do the left eye, but obviously he wanted me to come to terms with everything, and to allow everything to settle down on the right side. At this point, the left eye had rejected the cornea they'd replaced five years earlier.'

Overnight, Nina went from reading a newspaper without aids to once again being almost blind; her right eye obliterated and the left eye providing only the most basic, hazy vision. 'I was left with that sliver of vision to cope with.' Thinking back to that window of time with good vision, she says, 'It didn't feel very long. Just a year. I suppose, in my lifetime, it wasn't very long.'

A measure of Nina's resilience, she went back to work. But she found it difficult, and when she fell pregnant it all got too much. She decided to leave work and to concentrate on being a mother. Her son, Dylan, was born in 2010 and she got on with her life, using home adaptations such as the liquid-level indicators, those beeping devices clipped onto the coffee mugs. She describes her PenFriend, a device that, when held over the labels of household objects, plays a recorded message describing the contents of a container or the nature of an object.

Throughout this period, she continued to go to Moorfields every so often, to touch base with her surgeon. The vision in her remaining eye had continued to deteriorate further, and in

2016 yet another corneal graft was proposed, this time a partial graft on her left eye. 'I still had really good light perception. If Dylan was sat next to me on the sofa, I could see him, but I couldn't make out the details of his face. It was a risk, and it was my remaining eye. *Do we take the risk, or do we not?* So . . . we went for it!'

In contrast to previous grafts, this one was done under local, rather than general, anaesthesia. Nina recalls lying on the operating table with a plastic sheet over her head, the eye exposed. With the local anaesthetic, she could feel no pain, only tugging. 'It was very strange. Have you ever seen one of those films where the alien comes out with the light behind it, and you just see the shadow of the fuzzy alien coming through the door of the spaceship? That is what it was like, seeing him over me. It was very surreal. And they allowed me to choose the music in the room as well.' I ask her what she chose. 'The Stone Roses,' she laughs. Obviously.

In the recovery period, she had to lie flat on her back for a week. A bubble of air had been injected into the eye, maintaining pressure against the new tissue of the cornea. But once she'd recovered, she found there had been a huge improvement in her sight. 'It wasn't as good as before I'd lost my right eye. I could read again, but it had to be large print. I had to use magnification. But once I had magnified things, I could read them perfectly well.'

In the context of better vision, Nina decided to concentrate on what she wanted to do with her life. 'A lot of my life has been taken up by my sight problems, and every time I got to do something, or get a direction, I got taken back down by my sight.' She had studied design and art direction at university, and in the hiatus between losing her right eye and the most recent surgery, she had developed an interest in arts and crafts.

'It really helps psychologically. I just didn't think I could do anything else. I thought, *I'm not going to be able to use the computer to do the graphic design that I was trained in. I'm not going to be able to do anything like that now.* My auntie said, "Well, you know, you still are creative. That's what inspires you. That's what gives you that fire in your belly. So let's try something different. Still creative but something different." So, she took me to this bead-making class. I was like, "No, I can't do it." But she said, "Come on, we're going!" So we went and I made just a basic, simple necklace. And because it was tactile, while it was not really easy, I could do it, and I was really pleased with myself. And I was like, "Who needs eyes? I can still do something."' She laughs. 'And so I really got into that then, full steam – starting to make jewellery.'

With new-found enthusiasm, after regaining some vision, she decided to start her own business. She set up a creative café – quite amazing, considering that not only was she still healing from her surgery and coping with still quite limited vision, but was also a mother to her young son. 'I wanted to bring a space to people where they could come and get creative, and lose themselves in creativity. Forget about all the worries and the stresses and just lose themselves in that. We did really well. We had bookings every weekend for kids, craft parties . . . The kids absolutely loved it. We ran creative adult workshops in the evenings. And I was doing really well on social media, getting a great following. And I think that is when it all began to unravel.' I ask her what she means. Nina pauses for a second. 'Obviously, because I'd set the business up alone, I didn't have a business partner or anything. I had a lot of help from family – my family have always been a rock – but I was determined to do it myself. I think I just took too much on, because I was doing it all myself. I was doing the HR, the finance, the PR, the staffing, the training, the front-of-house. I was doing all of it. And I think it just got too much. And then one morning, a Wednesday, the twenty-ninth of August, 2018, I was getting my son ready for school . . .' For the first time since we have been talking, I hear a tremble in her voice. Her left eye glistens, and she gulps. I ask her if she wants to stop for a moment. She takes a deep breath. 'No, it's okay.' Another pause, before she carries on.

'We were having a bit of a clear-out in the house. We had bin bags everywhere. There was a broken TV on the dining table, and my briefcase was on the other side of the table, with the bank machine in. I was trying to get my son ready, I was trying to do the payroll for the staff because it was payday, and went to get the bank machine out of my briefcase. I was rushing, so instead of moving the bin bags out of the way, like a normal person would do, I climbed over, leaned down to get the machine out of the briefcase and basically headbutted the corner of the TV.' I wince, knowing what comes next. 'It burst the eyeball. It just burst a hole. And they told me that 95 per cent of the retina came out from the back of the eye through this hole. They told me that the corneal graft and the lens and everything at the front of the eye all stayed intact,' she smiles sadly. 'But they couldn't save the retina, only 5 per cent of it. I knew exactly what I'd done because it was the same as the right eye. Everything went black, and then I saw some lightning bolts, which I was later told was the retina detaching.' A small sob emanates from her.

In the aftermath of this tragic accident, Nina was lying on the hospital bed, recovering from the unsuccessful attempt to save her retina. Lying there, silent in her own darkness, she began to see – simple colours, waves of red and blue, muted not bright. She turned to her mother and husband, sitting next to her bed, and said, 'I can see something!' She could sense the two of them twitching with excitement, in the hope that some of her vision had been preserved: '"Oh my God, she says she can see something! She might be okay!"' An urgent call to the doctor was made, but when he arrived he told her it was just her imagination playing tricks on her. Her husband and mother bridled at this, insisting that he check. But while she could perceive the light of the doctor's torch shone into her eye, there was nothing else. 'But I knew I could see these colours!' Nina says.

A second operation was performed, to see if the surgeons could further salvage any of her retina. But when she came round after surgery, a brutal message was delivered. Nina remembers: 'They were very, very blunt about everything: "You have lost your sight. You'll never be able to see again." And I understand why doctors are like that. They have to be, I suppose. But there was definitely no emotion in it.' I ask her, from today's perspective, how she would have liked this news to be conveyed. 'I know they couldn't give me any hope, because there wasn't really any to give me. My sight had gone. I just think that the way that they delivered the message was just a bit harsh. I think you could be softer. Okay, it might be that you can't save my sight, and I understand that, 100 per cent, but I am still a human being with emotions that's going through a very difficult time. You know, I'm having to deal, and come to terms, with this.' A lesson in how not to deliver awful news.

Since that fateful day, Nina has been left in almost complete darkness – 'almost', because there is a tiny sliver of perception of the outside world. 'I still get a bit of light perception. It's on the left-hand side.' She points to an area just above the horizontal, a little to the left. 'If you imagine a black sheet, and then on the left there is a tiny little kidney-bean-shaped hole. And if you shine a torch behind the sheet, the light comes through that kidney bean.'

But despite her devastating visual loss, and the doctors' warnings, over the coming weeks and months Nina began to see. The waves of colour she had previously seen evolved into shapes, then patterns, and an ever-broader palette of colours. The blues and reds changed to yellows, purples and oranges, and eventually all the colours of the rainbow. 'They were geometric shapes. At points it could look like a kaleidoscope, at other times like a mosaic or a smashed tile.' As time went on, these images became more intricate, more complex. 'At first, I didn't really want to tell anybody, because obviously the first time I told someone, I was shot down. I just got more worried. *If they didn't believe me when it was just red and blue waves, how are they going to believe me now?*' For the first month, she told no one at all, not even her husband or mother. When they would ask if she was still seeing the colours, she would lie and say no. All the while, her vision filled with psychedelic colours, patterns and shapes. But as these images she was seeing rapidly progressed, she could no longer keep it to herself. 'Eventually, the shapes started turning into faces – cartoonish, animated. And then I began to see zombie faces; still cartoonish, but scary all the same – blood dripping from their eyes, and gnarly teeth. I knew they weren't real, but I was still scared. So I did speak to my mum. So then she told my husband, who told my auntie . . .' She chuckles; the perils of a large Irish family. Soon, everyone knew.

Nina clearly recognised that these visions weren't real. Initially, when she first mentioned the colours to a doctor in hospital, he suggested it might be light refracting or reflecting in the eyeball, so she assumed that it was indeed something to do with her eyes. But as the news of her hallucinations spread through the family, an army of relatives was deployed on the internet, and their searches quickly identified the hallucinations of Charles Bonnet syndrome.

For 250 years, the Republic of Geneva encompassed the south-west tip of the lake that bears its name. In 1798, precipitated by the execution by guillotine of Robespierre in Paris, the French army marched in, annexing the Republic for a short period of time. The Republic was briefly re-established for a couple of years before it became the twenty-second canton of the Swiss Confederation. But in the midst of a time of peace, in the mid-eighteenth century, one of the Republic's residents was a lawyer named Charles Bonnet. While a career in law paid the bills, Bonnet's true passion was the natural world. His fascination with insects and botany resulted in a litany of books and correspondence. But among his publications was a significant detour. In 1760, Bonnet wrote an essay on his grandfather's unusual experiences. At the age of ninety, Charles Lullin had undergone cataract surgery to both eyes and, much like Nina, had had an initial improvement before his vision began to fail. But what was taken away with one hand was given with the other. As Lullin's sight failed, he began to experience visions of another sort: vivid and detailed hallucinations of people, animals, carriages; others of buildings or changes in the tapestries hanging on the walls of his home. He knew that these were not real. It is all very reminiscent of Nina's experiences.

One of the hallmark features of the hallucinations of Charles Bonnet syndrome – a term coined not by Bonnet himself but by neurologists at the turn of the twentieth century – is the vision of small people, termed Lilliputian, after the islanders of Lilliput in Jonathan Swift's *Gulliver's Travels*. The visions are vividly described in a paper from the 1920s:

Small people, men or women of minute or slightly variable height; either above or accompanied by small animals or small objects all relatively proportionate in size, with the result that the individual must see a world such as created by Swift in *Gulliver*. These hallucinations are mobile, coloured, generally multiple. It is a veritable Lilliputian vision. Sometimes it is a theatre of small marionettes, scenes in miniature which appear to the eyes of the surprised patient. All this little world, clothed generally in bright colours, walks, runs, plays and works in relief and perspective; these microscopic visions give an impression of real life.

Bonnet described these vividly in his essay detailing his grandfather's visions. Lullin reported ladies or young girls immaculately dressed and coiffured, dancing or carrying objects in their hands or inverted tables on their heads. He described amazing details of diamond pendants, pearls, ribbons and other adornments. But Lullin, like Nina, also saw more elemental visions – whirling particles, clover patterns overlying everything in his vision, or the spinning spokes of a wheel.

Our understanding of the significance of Charles Bonnet syndrome has changed in the intervening 100 years. Initially denoting visual hallucinations in the elderly who are otherwise 'cognitively intact' (meaning without evidence of dementia or other neurological disease), it now refers to hallucinations in the context of any eye disease, in the knowledge that these hallucinations are not real. As Nina illustrates, being elderly is not a prerequisite. In fact, any eye disease can cause Charles Bonnet syndrome; you do not even need to be blind. The risk of these hallucinations increases with worsening vision, but you only need to have a drop in vision equivalent to about half-way down the optician's eye chart to experience these phenomena.

Of course, Charles Bonnet Syndrome – the brain's desire to see, even in the absence of vision – is not the only explanation of visual hallucinations. Visions of unreality are a common feature of psychosis. The major distinction is that, while in psychosis these hallucinations are indistinguishable from reality, in conditions other than psychosis people can clearly understand that they have no basis in fact, no matter how realistic or frightening.

And there are many other causes of these kinds of phenomena. Every week, I run sleep clinics and general neurology clinics at Guy's Hospital in London Bridge, and epilepsy clinics at St Thomas' Hospital in Westminster. They give me a change of scene, with different London landmarks as I walk between the two hospitals – the Shard and Tower Bridge at one site, the Houses of Parliament at the other. Occasionally, I get the River Bus to work, stepping off the catamaran at Embankment or at London Bridge pier. Despite living in this city for twenty-five years, at times like this I cannot shake the feeling that I am a tourist, particularly when the sky is clear and the Thames glistens in the sunlight, the usual murky brown of the water tinged blue.

In part, the separation of these clinics is logistical. The sleep and neurology departments are distinct, and often require slightly different tests and services. But, if I am totally honest, it happens to also facilitate my intellectual laziness. Sometimes, the continuous stream of individuals with a limited range of conditions contributes to my diagnostic focus, being either in the 'epilepsy zone' or the 'sleep zone'; a form of diagnostic blinkers, with its advantages and disadvantages. But in fact there is significant overlap between the world of epilepsy and sleep; similar techniques of studying the brain, similar medications, and sometimes similar manifestations. And for people with epilepsy, sleep is of fundamental importance. Poor sleep can precipitate seizures, small seizures can disrupt sleep, and anti-epileptic drugs may cause sleepiness or change the very nature of sleep.

It is for this reason that Susan sits before me today, in my sleep clinic. She has grey hair, cut short, her face lined by years in the sun; as we talk, I learn that she has spent some fifteen years living in the Caribbean, nursing her elderly mother-in-law, whose recent death precipitated Susan's return to the UK. I wax lyrical about the delights of the West Indies (my extensive knowledge of it is limited to films and TV) and how tough it must be to come back to the cold, colourless UK, but Susan quickly puts me firmly in my place. 'I couldn't wait to come back,' she says. 'It's nice for a holiday, but there is nothing to do. It is so boring!' She proceeds to tell me of the poverty, the crime and the rampant alcohol problems of the community she lived in, shattering my illusions of paradise.

The lack of proper medical care has been highly problematic there, and her return a couple of years ago has allowed her to seek treatment again. She has been referred to me by her epilepsy specialist at one of our sister hospitals. She has lived almost a lifetime of epilepsy, her existence punctuated by seizures – full stops, exclamation marks or semi-colons fragmenting the text of her life. Her specialist has noted that she has terrible sleep and, in an effort to improve her epilepsy, he has asked me to see her.

In the normal functioning brain, networks of neurones in the cerebral cortex – the grey matter that forms the outermost layer of the brain – communicate in a highly regulated manner. Everything is orderly, conversations between cells like whispers in the corridors of a huge bureaucratic organisation. But if something disrupts this normal functioning, be it damage to the cortex by a stroke, tumour or infection, or an alteration of how electrical signals are transmitted, through drugs or genetic mutations, the brakes on these impulses are taken off. The tightly controlled activity degenerates and the whispered conversations are replaced by deafening screaming and shouting. The corridors are filled with chaos, and huge swathes of neurones discharge in synchrony, causing dysfunction. Like the ripples on the surface of a pond provoked when a stone is thrown, this disturbance of normal function spreads across the surface of the brain. Dysfunction moves from one part of the cortex to adjacent regions, until either the seizure peters out or the whole brain is affected, causing a generalised convulsion.

Now sixty-two years old, Susan has had seizures since the age of eight. Her medications have changed over the years. Some have had no effect at all; others caused side effects. One drug caused insomnia, but after switching to another, the insomnia very much persisted. At least there has been an improvement in her seizures. She has not had a full-blown convulsion – the dramatic collapses with violent shaking we usually associate with epilepsy – for more than a year. But her seizures continue, several days a week, sometimes several times a day. Not generalised convulsions, but more unusual events, evidence of uncontrolled electrical activity in limited parts of her brain. We discuss her sleep, but it is the nature of her seizures that fascinates me.

Susan recalls her first seizure as a child. She was in the local swimming pool with her sister, splashing about. A boy, being irritating in a boy-like way, had begun splashing water in her face. She closed her eyes, but remembers droplets of water landing on her eyelids, the fluorescent lighting of the pool perceptible with eyes closed. A sudden sensation of nausea followed, and she turned to her sister complaining of feeling sick. 'Then I don't remember any more. I just woke up in the hospital. They said I'd had a seizure.' I ask her if she knew what that meant, at her tender age, but she tells me she had no idea. The doctors talked to her parents, not her, and she simply sat in the room listening to the conversation.

Further convulsions followed. She would lose control of her bladder function, and her entire body would ache for days from the violent shaking. I ask her how it affected her childhood. She was one of nine siblings, the oldest thirteen years her senior, the youngest four years her junior. She was accompanied by a team of family bodyguards wherever she went. 'There were a lot of us, so I could go to the park with my sisters and brothers. We still went swimming, but they had to keep an eye on me.'

Over time, it became clear that there were particular triggers for Susan's seizures. Flashing lights or flickering in her vision; sitting on a train with the sunlight falling across her face, shadows of the rail-side poles simulating a pulsating light; the reflection of a television on the walls of the living room; a strobe light in the local disco at the age of sixteen, abruptly ending her clubbing days; even the visual stimulation of certain types of plants swaying in the breeze (plants with soft leaves, when the leaves move independent of each other); and, bizarrely, the transient change in light caused by closing her eyes during a sneeze.

Susan has photosensitive epilepsy, which means certain visual patterns or stimuli trigger abnormal electrical activity in the visual areas of her brain. Something about these areas of her brain means that the normal processing of visual information is unrestricted. Even the act of seeing can precipitate these seizures. What is normally highly regulated, dampened down, has lost its brake. Like a car suddenly jumping out of gear and rolling down a hill, accelerating out of control.

For most people with this type of epilepsy, the cause is genetic – a mutation or multiple minor genetic changes altering the function of those ubiquitous ion channels, making the entire nervous system more excitable than it should be. But investigations have shown that in Susan there is an alternative explanation. Detailed scans of her brain have identified some unusual findings.

The blueprint for our nervous system is formed when we are still small embryos inside our mother's womb. Roughly three to four weeks after fertilisation, a small crest of tissue forms, running from the top to the bottom of the embryo, slowly folding over itself to form a tube. This neural tube is the basis for our central nervous system, creating the brain and spinal cord over the coming weeks. But later in our development, the cells that eventually create our cerebral cortex also migrate from here. Their birthplace is on the inside of that neural tube, gradually being pulled through the substance of the brain tissue to come to their final position on the outside of the brain. But the scan of Susan's brain shows that something has gone not quite to plan in the back of her brain, in the occipital lobes, the area of the brain responsible for vision. Small islands of tissue, cells that should have made it to the brain's surface, remain suspended in the substance of the brain itself. Small nodules of what should have been cerebral cortex are floating deep under the surface. The normally carefully regulated organisation of this area of the brain is disrupted. The usual checks and balances on the cerebral cortex have been unsettled, leaving this brain area liable to a loss of control. Certain visual stimuli trigger abnormal electrical activity, and Susan's subsequent seizures.

But what really grabs my attention is Susan's description of the prelude to her previous convulsions. She describes her auras, the warning that a seizure is about to happen. 'I begin to see different coloured balls appearing in front of me. They then float across my eyes, and then they start to pulsate. They veer left. Then suddenly I just see everything go white, and I'm gone.' The next thing she realises is coming round, usually on the ground, aching throughout her body, wet from incontinence. I ask her about the coloured balls in a bit more detail. She reports circular shapes, different colours, invariably starting just to the upper left of her centre of vision. They begin to throb before tracking further left. Sometimes, a split second before losing consciousness, she feels her eyes being irrepressibly drawn to the left. Then all goes blank. 'I can't control that pull. And that is when I know a big seizure is going to happen.'

While the changes in her medication have brought her epilepsy under better control over the past year and she has not had any full-blown convulsions in a while now, the auras – essentially smaller seizures – continue unabated, although with some minor alterations. She tells me she is still having them on a regular basis, sometimes several times a week. But if the blobs do not throb or veer to the left, she knows she will be okay; the aura will pass after a few seconds, disappearing as quickly as it came. The medication has inhibited the seizure activity, limiting it to a small part of the cerebral cortex and preventing this unregulated electrical activity from spreading more widely, to the whole of the brain, staving off a generalised convulsion. She is describing a textbook occipital lobe seizure.

As with all epilepsy, the experience of it depends on where in the brain the abnormal electrical activity arises. And this is particularly the case for seizures arising from the visual areas of the brain. The most common manifestations of this type of epilepsy are what are termed 'elementary visual hallucinations', precisely what Susan is describing. Coloured, often multi-coloured, circular patterns, spots, circles or balls. Sometimes, as the seizure progresses, the components multiply and enlarge, beginning to move, as the electrical activity spreads across the primary visual cortex, the area of the surface of the brain that is the direct input for visual information, where crude sensory impressions are received. For anyone who has suffered from migraine, these symptoms may be a little familiar. It is common for people with migraine headaches to experience an aura, a symptom that may precede the headache. Indeed, the first warning I myself have that a migraine is going to arise is a little flickering in the extreme edges of my field of vision, initially so subtle that I am not sure if I am imagining it. As it develops, and with it the certainty that this is indeed a migraine aura, I find it increasingly difficult to see through this area of flickering, which gradually grows and spreads across my visual field. It is at that point that I reach for the ibuprofen, knowing what is to follow. Very occasionally, I can see wiggly lines or zigzags, always black and white, progressing across my vision, a harbinger of pain. And this aura of migraine, like that of epilepsy, is the spread of a change in electrical activity crawling across my visual cortex, activating and disrupting my vision at the level of the brain. But it is the nature of that electrical change that differs in migraine. Rather than the rapid, highly disorganised discharges of epilepsy, in migraine this is more controlled, spreading more slowly, activating the neurones in a different way. The slow burn of a candle wick versus a spark touching gunpowder.

In Susan's case, as the seizure spreads from the primary visual cortex it affects areas of the brain further forward, including regions involved in eye movement – causing the sensation of her eyes being pulled – before the whole of the brain is involved. But in epilepsy there are also other ways in which vision can be disturbed. The nature of these disturbances directly informs us of some of the organisational make-up of the processing of vision. The act of seeing is so much more than registering that an image has fallen on your retina. Imagine seeing a family member or a friend. You see her face atop her shoulders, in the doorway of your house. You recognise her, understand her relationship to you, remember what she means to you, recall the last time you saw her. In order to make sense of the eyes, the nose, the mouth in front of you, there are two crucial bits of processing required. The first is: what does it all mean? What is this object, what is its significance, what are its associations? You understand that it is a face, a face familiar to you, that it belongs to your mother, who you love. The second important bit of information is: where is this face? Is it close or is it far? Is it close enough for me to talk to, to hug? Is that face coming closer or going away? Do I need to open the door wide to welcome her, or shut it behind her?

Without the 'what' and the 'where', a face is meaningless, just a collection of features without any substance. So for normal vision, it is not just seeing that is important. The process of making sense of the visual world requires integration with other cognitive processes, other areas of the brain. Streams of information therefore need to percolate from the primary visual cortex, at the tip of the occipital lobe, above the nape of the neck, to the areas of the brain interpreting meaning and location. The 'what' pathway flows forward into the temporal lobes, areas of the brain important for ascribing meaning to visual and verbal cues, integrating with memories. Within the temporal lobes are areas responsible for recognition of visual objects, like faces, or other complex images. And depending on where seizures arise, as we move forward from the visual cortex to the temporal lobe, the nature of visual hallucinations triggered can also become more complex: people, animals, figures or scenes; sometimes familiar, at others frightening.

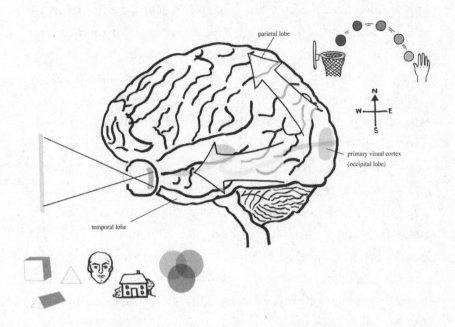

Figure 2. The 'what' and 'where' pathways of vision. Visual information is initially relayed to the primary visual cortex in the occipital
lobe, where conscious vision begins. From here, visual information
flows up towards the parietal lobe, which is involved in the integration of spatial orientation and movement (the 'where'), and forward
to the temporal lobe, where meaning is ascribed to visual objects,
colours, shapes and faces (the 'what').

Susan describes a second type of seizure, one that I have never
come across before. This has only developed in the past couple
of years. 'It's just arisen out of the blue. Each time the same.
It lasts twenty or thirty seconds, then disappears.' She begins
to describe these events, but they are difficult to explain and
equally difficult to comprehend. In keeping with other people
with epilepsy, sometimes the experiences themselves are
so beyond our scope of understanding that they are almost

impossible to articulate. She begins, 'It's like a negative. Like looking at a photograph. But if you are in front of me, I can see what is behind you.' She goes on to describe seeing a person standing in front of her, but also being able to see the scene immediately behind the person – the wall, a picture, or anything else in the background that should normally be obstructed by the person in view. Double vision can often cause people to feel they can see through objects, the two images misaligned, but Susan is adamant that she sees one of everything. I ask her if she thinks the colours are inverted, like a negative image, but she says no. 'It's like someone is translucent, but their image is superimposed on the background.' After some discussion, we agree that the best way to explain it would be two old projector slides, one on top of the other, or a double-exposure print. The image of the foreground and background superimposed, but without adequate explanation. Susan says, 'There is no explanation, logically, for me to be able to do that. I can just see through you!'

The rationale for Susan's inexplicable X-ray vision is problematic. Her vision in these moments seems to defy the laws of physics, a bending of light around a person standing in front of her. But there is indeed a possible explanation – that her seizure is disrupting the 'where' pathway. Somehow, this second type of seizure is interfering with her perception of visual space, of where everything is in relation to her world; a dysfunction of perception of where someone is in relation to the room, perhaps causing the perception of seeing both the person and the room, reconstructed from visual memory. The fundamental neurological basis of our map of the world, our location in it, and even where the various bits of our body are in relation to each other, resides in the parietal cortex, immediately adjacent to the occipital cortex, the visual centre of our brain. Interfere with these circuits, these streams of information from the occipital to the parietal lobes underlying our 'where' visual pathway, and you interfere with the knowledge of where visual objects are in space. Seizures arising in these regions may result in distortion of visual objects, with faces or people rendered grotesque, or the persistence of visual objects after they have gone, sometimes repeated over and over within the field of vision, or, at their most extreme, a tilting or complete inversion of the visual field. And so it seems that, with medication, another seizure type has come to the fore in Susan's case, another of these small islands of abnormal cortical tissue triggering small seizures disrupting the 'where' rather than the 'what'.

Susan's visual hallucinations, be they coloured blobs or distortions of visual space, illustrate something fundamental. Since seizures are a manifestation of overactivity of the cerebral cortex, the outer lining of the brain, her experiences clearly demonstrate that hallucinations originate from the cerebral cortex itself. The nature of the hallucinations is dependent on where precisely the seizures are originating from, but it is dysfunction, or overfunction, of the highly complex network of cells on the surface of our brains that is the basis of her visions. And this may provide some important insights into Nina's case too.

Nina recalls Charles Bonnet having previously been mentioned at one of her appointments. But no description of it, no explanation. At that point she had no idea that the problem was with her brain, not her eyes. Armed with the details gleaned from the internet by her family, she went back to hospital, demanding more information. 'I spoke to my consultant, and she was really nice and understanding, but really didn't give me any further information. "Yes, it's Charles Bonnet syndrome. I'll see if I can get you a leaflet for it." It was like, "You'll just have to go home and deal with it." It was a very dismissive experience.' Not someone to simply accept matters, Nina eventually managed to get in touch with a patient support group providing information and assistance for people with Charles Bonnet hallucinations.

I ask Nina if she is still experiencing the hallucinations. As we speak, she begins to simultaneously describe what she is seeing. 'So, imagine that black sheet with the little kidney bean cut out. It is like the kidney bean draws the shapes to it. So, flowing round the kidney bean at the moment there's a kind of blue sausage river.' She laughs. 'People must think I'm crazy! And then at the bottom of the bean, there is a kind of patchwork face. And then there are greens, blues and pinks, very muted.' As we speak she says, 'Now it's changed again. It's just constantly evolving.' I ask her if she ever sees little people, like those Charles Bonnet wrote about, but she does not. But she does see little cartoon characters, such as Bart Simpson or Minnie Mouse, occasionally Mickey Mouse waving his arms around. Her description reminds me of the watches prevalent in my childhood, Mickey's arms replaced by the hour and minute hand, rotating around the clock face. 'Sometimes, they are within a collage of shapes. It's kind of *Where's Wally?* – if you can spot him,' she chuckles, referring to those illustrations of thousands of little characters in which you have to find Wally in his red-and-white striped jersey.

Nina's current descriptions do not sound particularly scary or unpleasant. I ask her what it is about her experiences that she finds so distressing. She pauses again for a beat, considering her answer. She tells me that although she has now become completely accustomed to them – they are a constant in her waking life, present from the minute she wakes up until she goes to sleep – the intensity and form of her hallucinations fluctuates wildly. When she is having a bad day, tired or stressed, the colours become brighter, more vivid. And they become much more distracting. 'I find myself walking into walls or doorframes more, tripping more, because these visions are in my way.' And the zombie faces still come and go. 'It's definitely connected to my emotions. When I get sad, angry, anxious – any of those down emotions – that is when the zombie faces, devils, dogs, and so on, will start appearing. When I am happy and relaxed, it's just nice floaty colours and bubbles and unicorns.'

But what is the answer to Nina's unanswered question: why is this happening? What is it that causes the brain to generate these images? Is the need to see something, anything, so intense? Visual hallucinations are of course not peculiar to visual loss. They can be the result of a huge array of neurological and psychiatric conditions, such as schizophrenia, the effects of drug use, Parkinson's disease, Alzheimer's disease, even bereavement – and, as we have already seen, seizures and migraine. As I have already described, different areas and pathways within the brain are responsible for different aspects of visual processing. In fact, various areas of the cerebral cortex show a degree of specialisation beyond the 'what' and the 'where' of the visual world. And that specialisation is quite staggering, with some areas focused on processing colours, objects or textures, others on familiar faces, eye or mouth movements, body parts, object recognition, landscapes, or even letters or texts. Watch a yellow tennis ball flying towards you, and brain activity will increase in the shape, object, colour and motion areas of the cortex. Watch your partner smile at you, and the face-recognition and mouth-movement areas will light up. And, indeed, the same thing happens in the brains of people experiencing hallucinations, regardless of the underlying cause. If you are hallucinating the tennis ball or your partner's face, the same areas will show increased activity on brain-scanning, whether due to Charles Bonnet syndrome, schizophrenia or anything else.

It is perhaps easier to understand why changes in brain activity may be related to brain disease, with associated changes in both the structure of the brain substance and chemical transmitter systems. But it is less obvious why eye disease might cause something similar. However, at the heart of this is the so-called 'deafferentation theory' – that removal of inputs from an area of the brain leads to decreased inhibition of that area of cortex, causing hyperexcitability. Essentially, normal inputs prevent the brain from generating its own activity. Just as sensory deprivation in the context of torture causes people to hallucinate, visual deprivation through eye disease causes visual hallucinations. If a neurone has all inputs removed, it may die. But if any inputs are preserved, then that neurone or the environment it inhabits may adapt. New connections between the neurone and its neighbours sprout, and existing connections between neurones may adjust. Normally suppressed connections may increase in strength. Changes may also occur on either side of these connections between neurones, in the synapses, with increases in the pool of chemical transmitters on one side of the synapse and increased production of receptors on the other side of the gap. The net effect of these alterations is to trigger activity within that neurone that is without its normal inputs – on a larger scale, spontaneous activity of the cerebral cortex in the absence of information. The perception of vision in the absence of vision. And, as I will explain later, it is not just the sense of vision that is subject to deafferentation.

So it seems that, in Charles Bonnet syndrome at least, removal of sensory inputs predisposes to spontaneous activity in the visual parts of the brain, and it is the areas of the brain that are more active that define the nature of the hallucinations – whether they are simple or complex, their colours, shapes or the presence of

Lilliputian figures. But there remains an unanswered question as to why these Charles Bonnet hallucinations are so characteristic, particularly when it comes to those tiny figures elaborately dressed in historical costumes. This is yet to be fully understood. Perhaps it is due to certain highly specialised areas of the visual cortex being more vulnerable to deafferentation than others, influencing the nature of what Nina and others like her see – faces, figures, distorted in size or shape.

But there is another way to view Nina's hallucinations. In fact not just Nina's, but others too – those of people with psychosis, sensory deprivation and other disorders. And this explanation encompasses all our senses, not just vision, and ultimately comprises a fundamental aspect of our perception. We have already seen, from Beecher's descriptions of the war-wounded, that sensation is not simply a process of gathering information from the periphery and funnelling it to the brain, but that actually the brain can influence the data being captured. This is referred to as bottom-up and top-down processing, respectively. But this two-way flow of information is not limited to touch sensation, or even our senses in general; it is a feature of how every tenet of our nervous system works.

When it comes to understanding our world, there are three major flaws in the system that Nature has provided us with. The first is that the quantity of information that we are constantly bombarded with is simply too vast for our limited nervous systems to be able to process. When we perceive our world, it is like trying to stream an HD movie over a slow internet connection. The bandwidth is too narrow for all the data to be transmitted reliably.

The second issue is that we are essentially living in the past. Due to the make-up of our nerves, spinal cord and brain, and the connections between neurones – the synapses, which are reliant on the release of chemicals to send signals from one nerve cell to another – there is an inherent delay in our perception of the world. Consider standing on a tennis court at Wimbledon. Opposite you stands your opponent, repeatedly bouncing the ball, readying themselves to serve at match point. From the moment the ball leaves their racquet, you have about 400 milliseconds before it is likely to hit you in the face. When light hits your retina, it takes about 60 milliseconds for those signals to reach your primary visual cortex, at which point you vaguely perceive something, but by the time these signals have reached the other visual areas responsible for telling you the 'what' and the 'where' of that ball, it's probably more like 160–180 milliseconds. So if you are reliant purely on the signals coming in, by the time you actively perceive the ball in motion, it has already travelled several metres and is about to fly past you.

The third and final problem we have is the intrinsic ambiguity of any sensory information. Imagine seeing a red car not far in front of you. It would be safe to assume that it is indeed a car, several metres away. But based solely upon in the visual image falling on your retina, it could also be a small, model car just a few centimetres from your face.

These limitations of our nervous system are less obvious in day-to-day life. It is when we are faced with visual illusions, those in the pages of textbooks or the drawings of M. C. Escher, that some of these issues come to the fore. Take the simple block pattern that can be perceived as two black faces gazing at each other or a white vase; the line drawing of what appears to some to be a duck's head, to others a rabbit; or the picture of an old lady that can suddenly change into the side profile of a young girl. These are not mere curiosities, but illustrate the ambiguity that our perceptions have to deal with to make sense of the world.

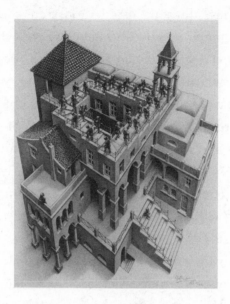

Figure 3. An example of M. C. Escher's staircases, *Ascending and Descending* (above), and the crone/girl illusion (below). Our brain resolves visual ambiguity by making predictions based upon what we see, related to our expectations. If you first see the crone, and finally make out the girl, it is then difficult to see the crone again.

But some of these illusions also demonstrate something more fundamental about ourselves. They give rise to the notion that the brain is not simply an absorber of information. It is a prediction machine. Our perception of the world is based upon predictions of how we expect our world to be, a necessary shortcut to deal with those three flaws, of data capacity, inherent delay and ambiguity. Let's go back to the picture of the old crone in Figure 3 – once you see the young woman hiding within it, it is difficult to go back to seeing the old lady. Then there is the illusion of a mask of a face constantly rotating through 360 degrees – as the mask turns and we see its reverse, a face turned inside out, we still see a normal face. Our brains have spent decades learning what faces look like; the nose and the chin being closer to us than the lips or the eyes. We expect to see a normal face, we predict a face to be there. Our perception is therefore what we know to be a face, rather than the inversion that we should be seeing.

Other illusions even demonstrate how we use prediction to compensate for living a few milliseconds in the past. This phenomenon can be illustrated very clearly in the form of an animation called the flash-lag illusion. A small bar rotates around an invisible centre point, like the second hand of a clock, flashing on the screen at predictable intervals, much like a stop–start animation giving the perception of movement. If an additional bar is flashed alongside the first bar, at the same angle, then the two bars look to be aligned as they rotate around the central point. But if the second bar appears only at unpredictable intervals, the two bars are perceived as misaligned, a kink in the hand of the clock. The predictable nature of the movement of the first bar means we are actually perceiving the bar to be where we expect it to be, while the unpredictable flashes of the second bar mean we can only see where it actually is, resulting in the two bars appearing to be set at an angle to each other.

Even on a cellular level, we can see evidence of some of the solutions that evolution has undertaken to deal with our flawed nature. The problem of bandwidth has long been recognised in the electronic world. In the world of television, one of the challenges has been to transmit the information detailing every pixel on a screen. But it has been known for some time that the colour of any pixel can often be predicted by the colour of the pixels next to it. For example, if a blue sky is on the TV screen, the likelihood is that one blue pixel is likely to be surrounded by other blue pixels. This suggests that there is a degree of redundancy in this information; that we do not necessarily need to transmit details of each pixel twenty or fifty times per second. What is more important is change; to communicate where the pixel colour has altered. This enables significant compression of the data being transmitted, like the zipping and unzipping of a large computer file sent by email, whereby redundant information is excised. It is change that we are interested in from a biological perspective, too. A static world around us is no threat, but as we move through it, or as it moves around us, this is what influences our survival – to perceive food or water, or the lion about to pounce. Our sensitivity to change is obvious in daily life, as is our ability to adapt to it. The heart-stopping cold of jumping into a lake slowly fades, and the stench of a room, apparent when we enter it, gradually becomes imperceptible, only brought back to our attention when someone else notices it.

Thus, there is a clear evolutionary pressure to detect shifts in our environment, but our nervous systems simply cannot sustain the bandwidth necessary to convey every single sensory input to our brains at every single moment. Evidence of the adaptation of our nervous system to cope exists even at the most basic level. Returning to the visual system, we know that the retina is chock-full of photoreceptors, with some 120 million rods and 6–7 million cones per eye. But when the photoreceptors are triggered, they do not transmit 250-odd million signals to our visual cortex, otherwise system overload would ensue. Instead, there is a class of cells in the retina called horizontal cells, which link groups of photoreceptors together. One of the functions of these cells is to detect relative signals from these groups of photoreceptors, essentially cancelling out similar signals and conveying messages to the brain only where there are differences between various photoreceptors rather than absolute values. A simple example of data compression.

By this point, you may be wondering what relevance this all has to Nina and other people with visual hallucinations. These limitations within our nervous systems that I have set out are essentially insoluble. Our brains are not limitless computers, able to process and draw meaning from our environment at every single moment of our lives. Shortcuts are necessary in order to facilitate our survival. And this shortcut comes in the form of prediction. The concept that the brain is a prediction machine, that our sensory inputs are interpreted in the context of our predictions of the world, is now firmly rooted in the world of cognitive and computational neuroscience. Within our brains, we have a model of the world as we understand it, based upon our previous experiences. For everyone on Earth this model is slightly different, based upon our genetics, upbringing and life experiences to date. The model is constantly being refined or adjusted, based upon the experiences of the day. In fact, as I have written about before, in *The Nocturnal Brain*, one hypothesis as to why we dream is that the act of dreaming represents the refinement of our internal model, while we are removed from our external world. A degree of prediction is vital to enable us to deal with the constraints of bandwidth, delay and ambiguity inherent to us. The predictive-coding model of our brains proposes that, at every level, there is a balance between incoming sensory inputs and this predictive process, the 'bottom-up' signals of sensation and the 'top-down' predictive signals feeding back on each other. Originally developed to explain aspects of vision, some theorise that this model could be extended to even higher levels of cognition and perception, including social interaction and belief systems.

Getting the balance right between the two streams of information running in opposite directions is a tricky business. We want to use our predictions to optimise our perception of the world, but if we over-rely on our predictive nature, we run the risk of creating the world in the image of our own expectations. If we over-rely on our sensory inputs, or under-rely on our predictions, however, we risk either being drowned in a sea of noise or being unable to overcome those limitations of our brains and nerves that I have described above. The balance is crucial, and it is at the crux of a unifying theory as to the nature of hallucinations. In Nina's case, the lack of sensory signals due to her blindness results in an over-dependence on her internal predictions, a relative imbalance causing an overweighting of her predictive brain, a perception of vision in the absence of seeing. There's a similar explanation for those undergoing sensory deprivation. But for the hallucinations of psychosis or underlying brain diseases such as Alzheimer's or Parkinson's, the internal predictive nature of our brains overrides those sensory inputs, also resulting in hallucinations. This view may also explain delusions – strongly held beliefs in the presence of clear evidence to the contrary – because the internal model of our world extends to all aspects of our understanding of our environment, not only our sensations. And this model of how we function may even extend to other areas. For example, people with autism often report 'sensory overload', a feeling of discomfort or inability to cope with too much sensory information, such as loud noises or a busy field of vision. Perhaps the strength of their internal world is weaker, limiting their ability to process sensory inputs.

As I wrap up my conversation with Nina, I begin to properly absorb the story that she has just told me. Now thirty-five years old, in the past three decades she has had almost countless promises of improved vision repeatedly dashed. Some of these incidents are more predictable – the rejection of any graft is a distinct possibility. But I cannot quite comprehend the last two incidents. Both times, it seems like fate has sought her out, accidents of infinitesimally small likelihood targeting her better eye. Catastrophic events; destiny conspiring against her. I ask her how she makes sense of all of this. She tells me she was never really spiritual before, brought up a Catholic in name alone. But with everything that has happened to her, she says, 'I honestly believe that this is where I'm meant to be. Here, sharing my story with the world. Creating an awareness. This is why I'm going through this journey, so that other people can hear it and learn from it, and hopefully help other people going through the same thing. It's got to be the reason for it. I honestly, truly believe that the universe wants me to be in this place.'

Her growing spirituality has also resulted in a leaning towards Buddhism and its principles. She has begun to meditate and has undertaken a course in massage therapy, something she enjoys. But she adds, 'When I'm massaging somebody, it's like I can see them. I know this might sound crazy, but I don't know if it is my third eye. When I am in contact with the other person, it is like I can physically see them on the bed in front of me. And the hallucinations go.' Her son enjoys having his feet massaged. Nina loves doing this, 'because it means I can see that little bit of him.' I am intrigued. I ask if she has ever massaged his face, to 'see' him, having not seen his face for several years, but she has not. She tells me, 'I don't want to be one of those blind people that say, "Let me touch your face,"' but says she has tried this once, with a friend. 'It sounds weird, but I could see the shape of her face.' It is vision of a sort, not through her eyes but through her fingertips.

In an effort to seek answers, and possible solutions, Nina has been put in touch with one of my colleagues, a world authority on Charles Bonnet syndrome based at a sister hospital in south London; it is in fact through him I have been put in contact with Nina. My colleague has raised the possibility of treating Nina with medication to dampen down her hallucinations. I have a final question for her. I ask, if she could be rid of the hallucinations, would she take that opportunity? She is uncertain. 'I ask myself that every day, especially on the bad days. But then I think, when they go, then what am I left with? Just darkness. At least now, I still have something to see.'

3

THE STENCH OF A ROSE

'Nothing revives the past so completely as a smell
that was once associated with it.'

Vladimir Nabokov, *Mary*

The worst stench I have ever experienced is seared upon my
memory, still gently smouldering some twenty-five years
later. I can smell it now as I consider it. It was 1995 and my
first month as a medical student on the hospital wards. The
preceding three years had been spent in lecture theatres,
tutorials, dissecting rooms and laboratories – anywhere but
face-to-face with patients. Perhaps we were not trustworthy
enough to actually meet the general public, too ignorant to be
let loose. Straight out of the blocks, my first 'firm' – the team
of consultants and junior doctors I was attached to – was in
vascular surgery. Fresh-faced, eager to learn and excited to be
meeting patients for the first time, I was quickly disappointed
to meet my mentors. The hospital that I was based at was, at
the time, still known to be a bastion of the medical establishment – old-fashioned, traditional and not very progressive.
The very essence of this was distilled in some of the consultant
surgeons on the firm, like Sir Lancelot Spratt from *Doctor in the
House* but without any obvious vestige of humour or humanity.

Their teaching styles were based on the established techniques of shouting, bullying and humiliation, and this ethos percolated down the firm structure, with registrars and house officers utilising similar teaching methodologies on us lowly students.

Vascular surgery is largely focused on the repair of diseases of the veins and arteries: stripping out varicose veins at one end of the spectrum, life-saving repair of burst aneurysms of the aorta (the main artery leaving the heart) at the other. And it was old-school surgery – blood spurting across the operating theatre in time with the heartbeat, tourniquets, amputations. In those days, before the widespread use of statins and before the beneficial effects of these drugs became evident, vascular disease caused by the furring up of the arteries was very common. The vascular ward was full of people with severe diabetes or who were heavy smokers, with clogged up arteries, leg ulcers and aneurysms. My first ward-round, a teaching round comprising the registrar and six medical students, was a tour through the ward visiting selected patients who provided particular learning opportunities. We drifted round the over-heated ward, new starched white coats over our shirts and ties, and at each patient the curtains would be pulled closed around us; seven medics in white coats in a confined space, standing over the poor patient as they were prodded and poked while we talked medicalese. The registrar had an air of menace about him, enjoying the game of intimidating these unknowing and inexperienced young students.

As we gathered around the last patient, even before the curtains were drawn round a faint hum of faeces and decay thickened the air. In the centre of the hospital bed lay a tiny sparrow of a woman, perhaps seventy years old but looking over a hundred. Her face was wizened, she had deep creases around her mouth and a waxy complexion. Her shock of white hair was tinged yellow at the front by months, years, decades of tobacco smoke, her teeth blackened, the index and middle fingers of her right hand stained brown with tar. As she looked around at us, it was clear from her bewildered expression that she was in the very latter stages of dementia, uncomprehending of this strange medical world. Closing the curtains around the bed, that previously faint hum grew into a strong stench. Already overheated, the closeness of eight people confined in that small space became almost unbearable. Without further ado, the registrar, a strange little half-smile on his face, whipped back the cotton sheet covering the poor woman, like a magician unveiling his latest trick in the expectation of applause. But he, too, got rather more than he bargained for. As the sheet came off, it exposed the woman's gangrenous leg – awaiting rescue or amputation – rendered black and purple, with sloughing green oozing from it, arterial disease starving the leg of oxygen-carrying blood. This was his big reveal, an effort to shock and revolt his charges – certainly enough to turn my stomach and those of my colleagues. But the unfortunate patient, in her demented state, had also been doubly incontinent. The bedsheet, her nightdress, her legs – all were smeared in a thick layer of dark faeces, and a puddle of urine had settled in the middle of the bed between her legs. The odours combined – human waste, infection, dead and dying tissue – filling our noses, our mouths, percolating into our clothes. It is the closest I have ever come to fainting on the wards. My colleagues and I could not escape fast enough. Even now, as I write, I can catch a dim sensory memory in the back of my throat.

If I'm being honest, I have an absolutely terrible autobiographical memory. A frequent conversation with my wife will consist of her talking about a social event we went to, a conversation we had, sometimes even a trip we made. She will be able to recite, almost word for word, a conversation from twenty years ago, while I will have virtually no memory of what she is talking about. A short way into the conversation, I will have to admit to not being able to remember. But that ward round, that patient, that incident – I remember every single aspect of it: the face of the poor woman, the colour of the walls, the look on the faces of the other students, the smirk of the registrar transforming into horror. But most of all, that viscosity of foul air as it filled my nostrils, sank into my lungs and infiltrated my head; that smell, that odour of death, that disgust in the pit of my stomach. It is indelible, carved in stone.

Among all the senses, smell and taste are outliers, both in their nature and in our understanding of them. At the most basic level, compared with the other senses, we sample something very different about our environment through our nose and mouth. Vision, hearing and touch convert energy into experiences, be that electromagnetic radiation or mechanical energy. Taste and smell, however, inform us of the chemical milieu that bathes us, or that we ingest. They are the sampling of molecules in the air that we breathe, or in the food that we eat – a primitive, primordial experience of the world, shared with even the simplest of unicellular organisms, such as amoebae; an insight into the chemical soup we inhabit, reliant on sensing tentacles reaching out into the world, touching substances around us that might indicate food, danger or a mate.

Unlike the other senses, taste and smell are less granular, less informative about our environment, unable to easily localise, to identify, the source of a smell or taste. While we have two nostrils just as we have two eyes, two ears and two hands, the sensory organs within our noses are barely able to discriminate where a smell is coming from. We can move our noses into likely locations as we look for an increasing intensity of smell, but this is very different from our acute ability to see depth and distance, to locate the source of a sound, or to feel intricate details with our fingertips. Similarly, taste floods the mouth, but does not give us further refinement about its source.

Our understanding of these two senses has been hampered by a perceived lack of relative importance, for life in general and also in the context of clinical disorders. They are at the bottom of the football league of the senses. Blindness, deafness or numbness have so much more of an immediate impact on the functions of life, the fundamentals of living, than loss of taste or smell. After all, is it the end of the world if we can't taste an apple or smell a rose? But this alone is not responsible for the relative paucity of knowledge of taste and smell. The scientific study of these senses has been limited by something a little more prosaic. We can easily quantify and qualify light, sound or objects we touch, according to location, wavelength, frequency, intensity or pressure. But how do we do this for odours or flavours? Even when it comes to our language sur-rounding these senses, they are outliers. It is almost as if no interpretation is needed, or indeed possible. We have extensive vocabulary to describe sights, sounds and the feel of things, but the range of language for smell and taste is limited to what other things they smell or taste like. We describe smells in the context of our experiences, memories poking up from the depths of our minds. We express smells and tastes in terms of their meaning to us individually, unable to break them down into their constituent parts.

And this highlights something extremely important. Remember Paul and his inability to feel pain, or Nina and her vision in the absence of sight? These disorders of the nervous system illustrate the schism between the real world and how we experience it. While under normal circumstances, that disconnect between the real world and our perception of it is much more difficult to truly comprehend, when it comes to these two chemical senses – smell and taste – the clash between the objective and subjective is, for all of us, more intuitive. What we see or what we hear is directly attributable – at least for the most part – to the physics of our environment: the fact that I can see an apple in front of me tells me that there is a red sphere sitting on the desk a few feet away. But when it comes to smell and taste, no one would argue that these senses tell us something qualitative about the substances we are encountering. When we taste vanilla ice cream, does that tell us about the specific chemical structure of vanillin, the primary compound that gives it that particular flavour? When we smell perfume, does our experience in any way relate to the physical properties of the cocktail of airborne molecules entering our noses? No – our experiences of the ice cream or perfume are merely shorthand, our understanding of our chemical environment reduced to a taste or smell, a label given by the brain.

If we are simply sampling chemicals in the air or in our mouths, we might expect these senses to at least tell us something vague about the molecular structures of these chemicals. Indeed that is sometimes the case. Similar molecules may have similar odours: all amines smell of something roasting, fatty acids smell rancid, and aldehydes often smell of plants, like cut grass or leaves. But this is definitely not always the case. Sometimes similar molecules may smell totally different. Sometimes even the same molecule may smell totally different when it has a different 3D structure (so-called enantiomeric compounds), for example one version of a particular molecule smelling of tropical fruit, the other version smelling of rubber.

Few examples demonstrate the illusory nature of our chemical senses better than *Synsepalum dulcificum,* otherwise known as the miracle-berry plant. Native to West Africa, this shrub grows a few metres high and produces 2cm red fruit shaped a little like olives or acorns – so far, it may sound rather unassuming and not very miraculous. But the miracle happens when you chew the flesh of the berry. On its own, it is tangy and very mildly sweet, but nothing much to write home about. But after you've eaten the berry, if you then bite a lemon or a lime, even drink some vinegar, your mouth will be flooded with sweetness, an explosion of the intense taste of sugar. The sweet taste will linger for several minutes, until the active compound of the berry, a substance known as miraculin, is washed away by your saliva. On its own, miraculin does nothing. In the normal setting of the mouth, with neutral saliva, the miraculin binds to sweet receptors and simply blocks them, not really stimulating a sensation of sweet. But when the mouth becomes acid, when saliva mixes with something sour, the miraculin binds with salivary proteins and changes its structure, suddenly making it able to trigger these sweet receptors rather than blocking them. A rather remarkable illusion of taste, it is a potent illustration of the loose nature of the relationship between the physical world and our experience of it.

'I would have people make jokes of it, just laugh at me, basically. I'd be sitting there with a handkerchief over my nose, trying to stop any smells, trying to stop breathing, really. I'd breathe through my mouth instead of my nose to stop having to smell that smell, and people would just make fun of me. I even had somebody say, "Well, at least you're not deaf or blind."' The bitterness is evident in Joanne's voice as she tells me of her experiences of the past five years – bitterness at the lack of recognition of the impact of her condition, bitterness at the all-pervading effects of it.

Joanne's problems started innocuously enough. A simple head-cold, in her mid-forties, back in 2015; the kind of cold that is familiar to anyone in the UK, an intermittent companion of autumn, winter and, for that matter, spring. Joanne, based in the far north-east, in Tyneside where the biting wind whips in off the North Sea, would also without doubt have been no stranger to a cold. But this one lingered for weeks, causing some chronic sinus issues, and ultimately Joanne needed a course of antibiotics to shift it. Her cold settled and she thought nothing more of it – until a few weeks later, when she noticed something strange. 'I started to notice a really bad, distorted smell. I couldn't put my finger on it. It could be between rotting flesh or sewage. Really a foul, putrid smell.' As time went on, Joanne found that this stench became all-pervasive, all-consuming. 'It just got worse and worse. It was every second of the day.' But it was not only her sense of smell that was affected. The reek of decay permeated her food as well. 'Everything tasted either like a chemical taste or as if it had gone off, as if it were mouldy or rotten.'

Specific smells seemed to heighten the unpleasantness, and the range of triggers was not at all conducive to normal life. 'If I came in contact with cigarette smoke, cooked food, coffee, even fabric softener or perfumes, that putrid smell was just heightened twenty-fold.' The smell of mint in toothpaste induced gagging, leading her to seek out flavourless toothpaste. Going into work, surrounded by the perfumes and aftershaves of her colleagues, became intolerable. She went off sick for several months. Even family life became fraught. A Sunday lunch with her partner and her sister's family would become a trial. The smell of food, the smoke of the coal fire, people's personal scents, all became quickly overwhelming. 'So I just escaped outside. But you're surrounded by smells outside as well. Just by going to the shops, you'll pass people smoking a cigarette. I would run a mile in the opposite direction. Even the smell of cut grass – I couldn't stand it.'

All this quickly took its toll. Joanne found her new reality extremely distressing. 'I just wanted to sleep all the time, because that was my only escape. I wanted to hibernate. I didn't want to go out, socialise, go to work. I literally just wanted to sleep.'

Like the miracle berry, Joanne's sensory world had become illusory. The perpetual odour of death, putrefaction, decay – an illusion of smell, horrific rather than miraculous. Joanne's distress was compounded by a lack of help or understanding. Her first visit to her general practitioner was met with a blank look. 'My GP hadn't heard of it at all,' Joanne recalls. She was given various nasal sprays and medications, to no avail. I ask Joanne if her family thought perhaps that the problem may be psychological – I have previously come across patients so depressed as to become psychotic, and to have delusions of the world around them rotting or dying. 'I think they understood it was biological, but they just didn't know how to help.' As is so frequently the case, 'Dr Google' gave an answer of sorts. It gave her condition a name. 'I had never heard of it, and when I challenged my GP, they hadn't heard of it either.' That name is parosmia.

The simple act of inhaling. A slow, deep drawing in of breath, with mouth closed. The subtle narrowing of the nostrils, the vibration of turbulent air tickling the nose. An unconscious act, the simple attendance to life itself – yet simultaneously, and equally unconsciously, an act of vigilance, of sampling the external world, detecting danger, seeking food, family, a potential mate. The whole world studied through a single breath. And underlying this examination of the universe is the only point where the central nervous system meets the outside world, where our brain reaches out beyond the confines of our own bodies. When it comes to all our other senses, the brain is guarded, with gatekeepers fronting a defence like doormen to an exclusive club; peripheral nerves and sensory organs, such as the ear or eyeball, keep the world at arm's length. But when it comes to smell, our brain lets its guard down, its tentacles reaching into external space, directly grasping at the air. At medical school, among the first pieces of anatomy we learn about are the cranial nerves, the twelve pairs of nerves emerging directly from the brain, innervating various parts of the head, neck and trunk. They are responsible for head sensation, eye movement, vision, even the function of the stomach. But the first pair of these nerves, the olfactory nerves, are unlike all the others. Like the optic nerves, they arise from the cerebrum itself, not the brainstem, but unlike the optic nerves, almost all their path remains within the cranium itself, and their fibres come into direct contact with the exterior environment. Fibres of the olfactory nerves float free in the mucosa of the nasal passages before piercing through the colander-like cribriform plate of the skull, the bony structure constituting the roof of the nasal passages. But the nerve itself derives from the brain; it is basically the brain reaching outside the skull into the world.

The details of the olfactory nerve are staggering. In each nostril, within the roof and septum of each passage, a small area of mucosal tissue, only 2.5cm^2 in area, is dedicated to smell. Within this tiny area, the size of a postage stamp, are located some 6–10 million sensory neurones, each waiting to encounter chemicals breathed in.

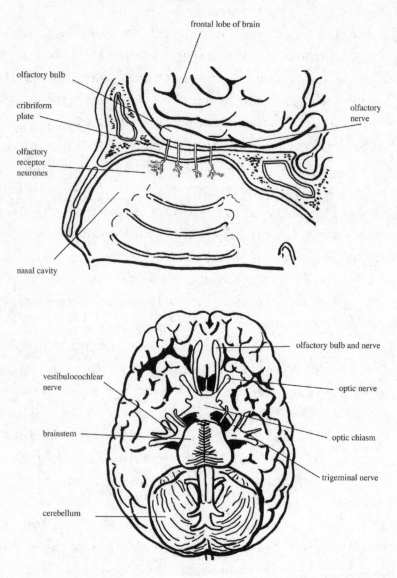

Figure 4. Above: cross-section of the nasal cavity. The olfactory receptor neurones pierce the bony cribriform plate, then form the olfactory bulb and transmit smell inputs to the brain via the olfactory nerve. Below: view of the olfactory nerve and the brain from beneath. The olfactory nerve sits on the underside of the frontal lobe of the brain. Other cranial nerves are visible, including the optic nerve (vision), trigeminal nerve (facial sensation), and vestibulocochlear nerve (hearing and balance).

Unlike many other parts of the nervous system, these olfactory neurones are constantly regenerating, lasting on average a month or two before being replaced by continuously replicating stem cells. However, exactly how these millions of neurones – like filtering sea anemones with arms outstretched, waiting for the current to provide prey – create a sense of smell remained largely a mystery until just a few decades ago. In the early 1990s, the Nobel prize-winning work by Linda Buck and Richard Axel identified a large family of genes that are responsible for detecting the volatile airborne chemicals dissolved in the lining of the mucosa – otherwise known as olfactory receptor genes. In mice, there are some 1,100 different olfactory receptor genes, while we humans have a paltry 350–370. Each sensory neurone expresses only one type of olfactory receptor gene, implying that each neurone detects only one chemical as a scent. But if that is the case, then how can we detect almost countless odours?

Subsequent work has uncovered a rather complex picture – a process that enables us to distinguish thousands of smells using a limited array of detectors. It is similar to the three colour-receptors in our retinas that permit us to detect a rainbow of colours through combining signals, but much more complicated. In contrast to the visual system, which has only one sort of input – light, albeit at various different wavelengths – the olfactory system has inputs in the form of chemical structures, unrelated molecules of different shapes and sizes. But rather than three types of detector, as with vision, the olfactory system, with its heterogeneous inputs, needs hundreds. Yet ultimately, there are parallels between the perception of colour from three different light receptors and the abstraction of smell from 370 different olfactory receptors.

While each individual sensory neurone does indeed express only one type of chemical receptor, any one odorant may stimulate several different receptors to a greater or lesser degree. Furthermore, each receptor can detect several related molecules. In essence, each chemical has its own signature, stimulating different populations of neurones to different degrees. And it is the summation of these signatures – multiple volatile compounds comprising the scent of a rose or the unpleasantness of soured milk – that constitute our own perception of smell.

As we begin to understand the gulf between chemical reality and our perception of our olfactory milieu (and the mechanisms that underlie it), it becomes apparent what a precarious system this may ultimately be. Firstly, there are the anatomical vulnerabilities. These neurones are in direct contact with the outside world – a very unusual occurrence. They are therefore at high risk – risk of damage from infection, inflammation or trauma. A simple cold, inflammation of the nasal passages, a blow to the nose – all of these may result in killing off these neurones. But their location in relation to other structures in the skull makes them especially vulnerable to injury. These nerve fibres must penetrate the skull, and do so via those small perforations in the cribriform plate. Any trauma to the head can result in small shifts in the internal organs of the skull, causing a shearing of these neurones as they pass through the bone. In fact, about 7 per cent of people with any head injury experience smell impairment, either through damage to these fibres or to the brain's smell-areas themselves, or through nasal injury blocking the passage of air to the olfactory receptors.

Secondly, since the entire system is based on each odorant generating a unique pattern of stimulation, anything that upsets the delicate balance between the inputs and outputs of the various receptors and neurones may fundamentally alter this sense of smell. Depending on the proportion of neurones damaged, and which particular ones are affected, the result may be a mild, moderate or total loss of smell. But it may also cause an alteration in smell, like Joanne's – parosmia, where the smell of everything is altered, where all is tainted by the odour of something else. Once again, there are parallels with vision, where the red, green and blue cone-receptors integrate wavelength information to permit perception of colour; lose one type of receptor and you become 'colour-blind' – your world of colour takes on a new hue. Most of us are aware of the genetic colour-blindness caused by an absence of one type of receptor at birth. But if the nerve, or indeed the retina, is damaged by something later on in life, for example certain antibiotics used in the treatment of tuberculosis, this may result in a similar phenomenon.

I sometimes see patients with seizures arising from part of the brain near the primary olfactory cortex, where smell is consciously experienced. In the prelude to their seizures, during the aura, they will often notice an unpleasant smell, of burning rubber or some other foul odour, as the uncontrolled electrical activity stimulates these smell centres of the brain. Clearly, then, this sort of phenomenon can arise from the brain itself, albeit very intermittently in the context of seizures. But parosmia is almost always seen in individuals whose sense of smell is already impaired more generally, adding weight to the view that it is the destruction or disruption of the olfactory neurones that lies at the heart of this condition, rather than anything at the level of the cerebral cortex. The description of the distortion of smell as being rotten, foul or sewage-like is almost universal, and most say that particular smells like petrol, tobacco, perfume or fruit tend to trigger their parosmia. The underlying cause is most frequently a bad cold, chronic sinus disease or head trauma, though it remains unknown in up to a quarter of people. In many respects, Joanne is a textbook parosmia sufferer.

One other very significant feature of parosmia, and of other disorders of smell or taste, is depression. More than 50 per cent of sufferers experience depressive symptoms as a result. In that respect, too, Joanne is no exception. While she does not state it explicitly, as I talk to her it is clear that her condition has had a massive effect on her mood. She readily admits to having wanted to do nothing but sleep, an escape from the torture of decay pervading every aspect of her waking life. 'You just can't even think about living a life when you're constantly smelling a putrid smell, tasting a putrid taste,' she tells me. She couldn't work, couldn't socialise, could scarcely bear to be in the same room as family members; the smells associated with other people heightened the foetid smell eating away at her, making life intolerable. But there were also other factors that contributed to her general despair: the duration of time it took to get a diagnosis, and the general lack of knowledge about her condition by medics and others around her; the merry-go-round of treatments without success; the anxiety caused not only by the parosmia, but also by the loss of certain aspects of her sense of smell – 'I couldn't detect gas or smoke. I could smell something, but it just smelled the same as any other smells. It was very worrying.' Even after she had been given a diagnosis, there was the uncertainty of the future. She was informed that there was no cure, and no one could tell her if her sense of smell would ever normalise. 'I became very angry and reserved, and I just didn't want to be around anybody,' she recalls. 'It's very difficult for my family to understand what I was going through simply from my description. It's really difficult for somebody not going through this to understand it at all, what it must be like.' Eventually her GP prescribed her a course of anti-depressants, but she remembers them having virtually no impact.

It is easy to understand why going through what Joanne was experiencing may cause depression; to understand the impact on family, social and working life, and on the desire to venture outside – in fact, to engage with the outside world at all. But there may be something else going on. Perhaps there is an alternative explanation for this mood disturbance, or at least another factor. As we have already seen, smell is an outlier when it comes to our senses. The pathways between its sense organs and our brains are more direct, less filtered. There is no peripheral nerve component; it is as if the brain itself is reaching out into our environment. For the other senses, the pathways to the cerebral cortex are lengthier and more convoluted, down nerves, through the spinal cord, through the brainstem or via various relay stations, clusters of neurones termed 'nuclei'. In contrast, for smell, this is not the case at all. Nerve fibres from the olfactory bulb – the bit of the olfactory nerve immediately above the cribriform plate – project directly into the olfactory cortex, the centre of our smell experience. But this primary olfactory cortex is surprisingly widespread within the brain, and consists not just of an area involved in the discrimination of smells, but also regions of the limbic system, the areas of the brain involved in processing emotion. It seems that these areas of the brain are responsible for representing pleasurable or unpleasant responses to smells; the amygdala, a key region of the limbic system, lights up on brain scans when people are exposed to horrible smells.

Beyond this primary olfactory cortex, the smell-inputs then project more widely to other areas of the brain. And when you look at which areas these are, there is a striking overlap between the regions involved in olfaction and those involved in depression. But these links between smell and mood are not solely anatomical. There appears to be a reciprocal association between smell and depression. People with clinical depression have poorer smell function than control subjects, and those with poor smell have symptoms of depression that worsen with the degree of smell impairment. Indeed, the volume of tissue that constitutes the olfactory bulb tends to be smaller in depressed patients. Even in animals, there are strong links. If the olfactory bulb is damaged in rodents, this results in changes to immunity, hormones and chemical levels similar to those seen in depressed humans, with changes in the concentration of dopamine and serotonin within the brain. The rodents' behaviour changes, too – again, more in keeping with depressive behaviour in humans.

And so it may be that, beyond the direct consequences of the parosmia on Joanne's lifestyle, there may be more fundamental reasons as to why she became angry, withdrawn and irritable; perhaps it is a direct effect on her brain. But this strong relationship between smell and emotion raises further questions: why should this relationship exist in the first place?

I am drawn back to the stench of the ward over twenty-five years ago, when the surgical registrar whipped back the patient's covers, revealing rotting, purulent flesh, urine and faeces. It fills my nostrils even now, and I can vividly remember the sudden sense of horror and revulsion in the pit of my stomach. I can vividly see the poor woman's expression as she looked at our faces, not understanding where she was, who we were, and what we were all doing there. Her face, that foetid smell, those emotions: they are etched on my memory, unfaded by the passage of time. While hopefully few readers will have encountered this specific scenario, almost all of us will understand the central tenet of this experience: a particular scent sparking the memory of an intense, long-forgotten moment, an emotion or scene, the ability of an olfactory trigger to draw us back to a specific time or place, often associated with a particularly positive or negative emotion. The smell of a cake baking in the oven, the perfume or aftershave of a long-lost love. Or, as my wife recently found, on a tour of a prospective school for our daughter, the smell of the school corridor so strongly evoked unpleasant memories of her own school that she took an instant dislike to the one we were visiting, instantly ruling it out. And then, of course, there is Marcel Proust, who is quoted in almost every scientific article on this subject: 'And suddenly the memory returns. The taste was that of the little crumb of madeleine . . . The sight of the little madeleine had recalled nothing to my mind before I tasted it.' For Proust, it was the flavour (not just smell) of the small cake that took him back to the past. A less unpleasant example than my own.

It is smell, then, more than any other sense, that seems to be evocative of emotional states and related events. While other stimuli, whether musical, tactile or verbal, are as good at conjuring memories of events, those recollections triggered by smells are consistently more emotional events. In fact, experiments have shown that emotional state may actually strengthen the link between smell and memory. In one experiment, pre-exam students were given word-based memory tasks, some with an ambient smell and others without. Those accompanied by an odour did slightly better, but this effect was strongest in those students who were highly anxious during learning.

So, it seems that not only is there an overlap between olfaction and emotion, but also olfaction and emotional memory. Those regions of the brain involved in olfaction and emotional processing also have a strong role in memory. The amygdala, with its direct inputs from the olfactory nerve, is fundamental to the experience of fear and 'fear conditioning', the process that pairs a previously neutral event (like the sound of a car braking) with fear (in someone traumatised by a serious car crash). This process may be deeply unhelpful in individuals with post-traumatic stress disorder, in whom the smells of blood, petrol or, in the case of Vietnam veterans, napalm, may precipitate a massively exaggerated fear-response associated with awful memories. But this process may also be crucial to survival. Consider a mouse's instinctive fear-response to the smell of a cat; it would be potentially very useful to be fearful at the smell of cat litter or cat food too. This sort of effect would also be helpful for humans. Equally, the entorhinal cortex, another region that constitutes the primary olfactory cortex, is closely linked to the hippocampus, a region of the brain fundamental to new learning and memory. A pleasure response to a particular smell may also be important from an evolutionary perspective, causing organisms to seek out smells that facilitate eating, safety or mating.

This tendency for smell to trigger memories may have some far-reaching implications. Studies have shown that odour-induced positive-emotion memories have a beneficial effect on the immune system and inflammation, as well as altering brain activity beyond just the olfactory regions. The 'simple' act of perceiving chemicals in the air – the binding of molecules to the chemical receptors in the olfactory epithelium – underlies a neurological and psychological system that has consequences for a multitude of cognitive, behavioural and physical processes. Smell, despite at its core representing a simple chemical detection system, is anything but simple.

Now, some five years after her parosmia devastated her life, Joanne has made a significant recovery. 'I would say I am about 80 per cent better. I don't smell a putrid smell any more. I don't have normal smell; there is still a distortion there. I still don't like a lot of perfumes.' Her path back to some semblance of smell normality has been long and convoluted. She eventually found an ear, nose and throat surgeon with a specialist interest in smell disorders, something surprisingly difficult to find. She was given an experimental treatment, a drug usually used in asthma, called theophylline. The hypothesis is that this drug may boost factors in the nasal mucosa that drive olfactory neurones to replicate, thus accelerating the normal process of replacement of these cells. If the underlying cause of Joanne's parosmia is the loss or damage to a sub-group of these neurones, then promoting regrowth should indeed help.

Whether through coincidence, placebo or the direct effects of this drug, Joanne tells me, 'I took [this tablet and] more or less within one or two weeks, it suppressed that putrid smell.' Joanne also ended up taking gabapentin, a drug initially designed for epilepsy but now used in a variety of neurological contexts to dampen down neuronal activity. She says, 'The combination got me to a much better place; to be able to cope with things. I do understand that the olfactory bulb in your nose does regenerate sometimes, so it could be that a combination of the two drugs has helped to suppress those smells until the olfactory bulb has regenerated.' She has now been off both drugs for about a year. Her initial fear of the parosmia returning as the drugs were withdrawn has not been realised and she continues to make small improvements despite being drug-free. She has already started a treatment called smell retraining. A couple of times per day, she holds a series of strong-smelling oils under her nose. The aim is to essentially retrain the nervous system to smell again. How it works is unclear, but it is presumed to stimulate growth of the olfactory receptor neurones, and may cause the reorganisation of pathways within the olfactory bulb or even the brain itself.

'I definitely do feel I'm making progress. When I was first diagnosed, they said that if my sense of smell didn't come back within six months, then it probably never would. However, it's taken four years to get to where I am.' The rate of progress now is almost imperceptible, and Joanne realises that it takes many months before she notices any marked change. 'So I would say any improvements were miniscule. You couldn't say a percentage every month or every couple of months. It takes a long, long time before you can notice any change. Before, I was being seen every three to six months, but now they've agreed to see me in two years' time to see how I've progressed.'

At the start of this book, I asked you to play a game of sensory rankings, to order the senses in terms of their importance to you and to define those you would be most willing to sacrifice. My own league had the chemical senses of smell and taste at the bottom – ones I could live without. But listening to Joanne, it is readily apparent that what at first appears to be a simple case of alteration in smell perception has much wider implications. Smell is not simply the detection of chemicals in our environment, but has a fundamental role in our mood and our memory. It is not only a sense that informs us about our exterior world, but also influences our internal world. And, as I will explain later, smell has a crucial role in our very survival, influencing our choice of food and selection of mate, and enabling our wordless communication with others.

Importantly, though, as with the other senses, what we perceive through smell is nothing more than an illusion, our brains and our bodies ascribing meaning to chemicals in the air. Joanne and her parosmia, like the miracle berry, show how minor changes in the functioning our of bodies, in disease or normal health respectively, can so markedly alter our perception of reality. How we perceive smells, and our emotional response to them, is driven entirely by the need for survival. I think back to that elderly woman in her hospital bed; that revolting odour, the air thick with death and decay, bile at the back of my throat. If I were reliant on rotting tissue or faeces for my survival, as a maggot or dung beetle, perhaps, that thickness in the air would be as delightful and comforting as the smell of baking bread, as sweet as a rose.

4

ALL THE NICE GIRLS
LOVE A SAILOR

'Then God said
let there be sound
and divided the silence
wide enough for music
to be let in and it was a good groove'

Camisha L. Jones, 'Ode to My Hearing Aids'

As I stand there in the London mizzle, that peculiar combination of mist and drizzle, I think what a surreal experience this is. I am in a garden in Hampstead, the gentle hum of the traffic in the distance. The garden is somewhat overgrown, in a deliberate way, and by my feet is a small pond full of frogspawn – efforts to create a haven for wildlife, balanced by some design touches that hint at eccentricity. There are bird feeders all around, but also wonderful sculptural renditions of birds and other creatures, and a small garden gnome hiding in the undergrowth. And in the midst of the vegetation, on a small patch of turf, stands Bill Oddie, hand cupped behind his ear, listening for birdsong while intermittently tweeting – literally, not on a phone. Speaking of the garden, Bill says, 'We've been here for a long time. I've completely changed the garden to something – as my wife would put it – ludicrous.'

Bill is probably my earliest memory of British television. I remember being at a friend's house, having picked up enough English to vaguely communicate. On the television in the background I could see three men on a trandem (a three-seat bicycle) and various stop–go animation-type sequences. My English wasn't good enough to understand what was actually going on, but I remember my friend and his mother laughing as *The Goodies* played. But it is Bill's voice that I most recall from my childhood: slightly raspy and high-pitched – very distinctive; the voice of some of the characters in *Bananaman*, a Saturday morning cartoon familiar to the children of the 1980s. Later, I would occasionally hear him on various comedy panel shows on BBC Radio 4, and in the past couple of decades he has been ubiquitous as a presenter of natural history programmes on television. He is probably Britain's most famous birdwatcher or 'twitcher'. But right now, he is simulating the songs of various birds in a suburban garden in north London, as I look to the overcast sky. I close my eyes briefly, and it is as if I am listening to the familiar voice on the airwaves. I need to remind myself what it is that I am actually here for.

At the rear of the house, overlooking Bill's oasis of calm, is a conservatory. The room is full of musical instruments, a large drum kit taking centre stage. Hanging off the cymbals stand is a faded *Goodies* cap, the only visible marker of his long and varied career. We sit on sofas as we chat, a series of birds flitting in and out of the garden behind him.

'I've just gone down to Specsavers,' his voice drops to a whisper, 'to try some new hearing aids, because I can't hear properly any more.' For anyone, a deterioration in their hearing is of significance, but for Bill it is particularly relevant.

'My hearing is very important to me. As a birdwatcher,

you do more of your identification through hearing than you do seeing. Any birdwatcher will tell you, you hear something and then: "Hello, where's that?" And then you look for it. So hearing is absolutely vital. And I had a reputation of having not excessively good, but pretty good hearing.'

The sound frequencies of the songs or calls of different birds vary hugely, and Bill says he began to notice some difficulties in hearing the higher-pitched calls of certain species. 'There are certain key birds that, as you grow older, you find yourself saying, "I can't hear grasshopper warblers any more – can you?" There's a little club of us: "No, we can't hear those either."'

Bill recalls one specific moment of realisation that his hearing was waning. 'I live near Parliament Hill. We were up there with another couple of friends, and there were little birds migrating – a bird called a meadow pipit – and I could pick out their call a mile off, normally. But this awful moment occurred when I could visually see a couple flying over and I said to my friend, "Are they calling?" He said, "Yeah." And another one comes over. "Is *that* calling?" "Yeah." And that was it. It was horrible. I said, "Whoops, I can't hear meadow pipits anymore." Then I tested myself on a few of the other high-pitched birdsongs – sometimes in reality and sometimes with recordings – and I realized that it was the top frequency that had disappeared.'

Since that moment, some four or five years ago, Bill has noted a gradual deterioration. Goldcrests, redwings, other birds too: 'I just can't hear them any more.' And the effects have now spread beyond his twitching. 'It is not terrible, although my wife would say yes, it's awful. She gets so cross with me when she has to repeat everything. It's a very damaging process, isn't it? Anyone who has hearing loss: your partner will go barmy at some point.' He has also noticed an ever-increasing challenge to make out speech in a crowd or at a dinner table – 'burbly, burbly', as he refers to it. Most of the time, it is the clarity rather than the volume that he has trouble with. 'And sometimes – she doesn't like to hear it – but I can't always hear my wife.'

About two or three years ago, Bill decided that the time had come for hearing aids. His first attempt to resolve the problem was only somewhat successful. He found that the hearing aids helped a little, but did not normalise his ability to hear birds or his wife. 'The hearing aids were a sort of payment for a Specsavers advert I did – which I wrote for them, actually; an early one, before "You should have gone to Specsavers" – that one. It was a nice little ad,' – he digresses – 'quite a nice little film. It wasn't presented as an ad, just a bit of promotion. I played the birdsong, and then played what I could hear and what I couldn't hear.' This first experiment with hearing aids, however, ended rather abruptly. 'I dropped them one day in a theatre – not a good place to drop your hearing aids, crawling around amongst everybody. They were never to be seen again.' His recent trip back to Specsavers, on the morning we meet, is an effort to try once more. He has ordered some top-of-the-range replacement aids – 'At the moment, I'm contemplating a fairly large bill,' he chuckles.

I glance over at the sizeable drum kit taking pride of place in the room, with musical instruments scattered all around. I ask him if they are all his. It turns out that most of the collection is his daughter's, although the drum kit belongs to him. 'I play it quite quietly, if at all. It's not a loud noise, you know. I know a lot of musicians, and an awful lot of them have hearing problems, but then they used to have ten tons of equipment in their earhole!'

*

If asked to define sound, most of us would say it is a pressure wave, a vibration, passing through the air to our ears. This is of course correct, but the process of hearing that sound is something entirely different. It is all too easy to envisage hearing as being akin to recording a voice for a gramophone record, with the singer or speaker shouting into the horn to make the needle cut the groove onto a master record, or a microphone converting those vibrations into electrical signals. But that is not what hearing is. Rather, when we listen, what we hear is the result of the process of making sense of these pressure waves all around us, ascribing meaning to these tremblings of molecules. It is an early warning system, an awareness of what lies in wait immediately beyond our bodies or outside our field of vision. It is also an effective mode of communication. As the authors of the textbook *Auditory Neuroscience* state, 'Every time you talk to someone, you are effectively engaging in something that can only be described as telepathic activity, as you are effectively "beaming your thoughts into the other person's head," using as your medium a form of "invisible vibrations".'

At the core of this process of understanding our auditory world is that the sound an object makes is a function of its physical properties and location. When mechanical energy passes into any object, it will vibrate, with the frequency dependent on its physical characteristics such as its size, its stiffness, its shape. Many people will be familiar with the pure tones played to us through headphones during a hearing test – a single tone, artificial, devoid of any deviation in amplitude, frequency or harmonic, not decaying with time. In the real world, however, all these features inform us about the world; pure tones, meanwhile, are a unicorn, almost imaginary. Consider perhaps the simplest of noise generators, such as a string on a guitar. Depending on the thickness of the string, the material it is made of and the tension it is under, when plucked it will vibrate at a particular frequency, its 'resonant frequency'. But there is a reason why that string does not sound like those pure tones on the headphones. The string not only vibrates at its resonant frequency, but to some extent will also vibrate at multiples of that frequency. Therefore, rather than one single frequency, the vibrating string will generate multiple frequencies – all multiples of the resonant frequency, termed 'harmonics'. It produces not a pure, single tone, but a combination of tones. In addition, exactly where along the string it is plucked will influence the relative amplitudes of each of those frequencies, giving a slightly different sound.

A string is a one-dimensional object of course. Two- and three-dimensional objects may vibrate in more than one axis, and so may generate a number of different resonant frequencies and harmonics. The clanging of a steel manhole cover on the street sounds very different (and extremely displeasing) compared with Eric Clapton's teasing of his guitar.

So the frequency, or frequencies, an object makes tells us something about its size, its shape and its composition. But there is more information that can be gleaned from sound. The vibrations an object makes eventually die down, decaying into nothingness. This too is dependent on its composition. Compare a steel bar with a plank of wood dropped onto a concrete floor. Both will make a loud noise, but the ringing of steel will last several seconds, the clatter of the plank only a brief moment. In wood, the energy is dissipated much more quickly, the sound silencing rapidly, while the steel continues to vibrate. The nature of an object is so fundamental to the sound it makes that we can even distinguish between the sound of hot and cold water being poured into a cup. As water heats, its viscosity reduces, causing it to bubble more when poured, resulting in a higher-pitched sound. Try it at home, with your eyes closed – but ask someone else to do the pouring!

It is not just the source of the sound itself that tells us what we're hearing. The environment, too, can influence our perception of sound – the reverberations of sound in an empty room, with hard acoustic surfaces, versus a small furnished room, with carpet on the floor, curtains on the windows: a form of sonar, telling us almost as much about the passive world around us as the sound source itself.

Knowing what is making a sound is clearly important, but so is knowing where it is coming from. In this respect, sound may be more valuable than vision, which can only provide information if the object of interest is directly in front of us and it is not pitch dark. Understanding the location of a sound source may mean the difference between life and death – where a predator is, where food, water or a potential mate may be found. There are some important features that help. As sounds pass through air, higher frequencies dissipate more quickly due to friction in the air, while lower tones travel further. Stand next to a busy motorway and the higher pitches of motorbikes and car engines will be easily heard, but a mile away from the same road you may only hear a low-pitched rumble in the distance.

What we hear may therefore inform us about distance, but *how* we hear is even more important when it comes to knowing where a sound originates. Identifying the accurate location of a sound source ultimately comes down to anatomy – the shape and density of our heads, the distance between our two ears, and even the shapes of our external ears. The advantage of having two ears, set a small distance apart, with the head in between, is that sounds will be heard differently in each ear. A sound from the left will reach the left ear a fraction of a second earlier than the right. The margins we are talking about are tiny, a maximum of 700 microseconds, but our brains are sufficiently sensitive to pick up the difference. And that sound from the left will also be slightly louder in the left ear than the right, which will be sitting in the acoustic shadow of the head. The difference in loudness or sound level detected by each of the two ears will also depend on the wavelength of the sound, as well as its interaction with the folds and ridges of our external ear, providing further important data for our brains to compute direction. What's more, our ears are asymmetrical in every possible way, meaning that the quality of the sound changes subtly depending on the angle at which the sound wave reaches our head.

One can now begin to appreciate that the act of hearing is very different from simply capturing sound; it is highly complex, reliant on calculations based upon tiny differences in timing, volume and tone, with a temporal resolution that blows the other senses out of the water. A difference of less than 700 microseconds is sufficient to inform us about the direction of a sound – roughly a hundred times quicker than the interval between light hitting your retina and visual signals reaching your brain.

But there are two major logistical hurdles in this entire process of converting sound into meaning. The first of these does not at first glance appear to be a major one, but is not without its challenges. It is the problem of converting the physical vibrations into electrical impulses, the language of our nervous system. This involves overcoming physical, anatomical and computational obstacles. The sounds we are exposed to range from almost zero decibels to 120 decibels, the volume of really loud rock music. When put in these terms, it doesn't seem a problem. But think of it in terms of energy levels, and the loudest sound has 1,000 billion times more energy than the quietest audible sound – the equivalent of having the same set of scales to weigh an ant and the mass of a planet. We also need to be able to hear a wide spectrum of frequencies. For the human ear, this ranges from 20 to 20,000 Hertz, meaning cycles per second, and we need to be able to hear a multitude of frequencies simultaneously.

A further challenge is the transmission of sound into the depths of the body. The body is of course liquid-filled, yet, for land animals at least, sound transmits through air. For sound, crossing the barrier between air and water is problematic because the energy required to shift water is much greater than that required to shift air. Thus, a sound wave propagated in air is simply too weak to cause similar-sized vibrations in liquid, so when the sound reaches the boundary between air and liquid, most of its energy will simply be reflected back.

Even once sound has been transmitted into the body, there is the issue of how to convert sound into electrical signals inter-pretable by the brain. The ear is Nature's engineering solution to all these problems. The ultimate translator of mechanical energy to electrical is the cochlea, a snail-shell-shaped struc-ture measuring about 9mm by 5mm, the tiny link between us and the auditory world. It is essentially a coiled tube filled with liquid, about 3.5cm in length if unfurled. Along its length runs a membrane – the basilar membrane – that is stiff and narrow at one end, floppy and wide at the other. The precise structure of the cochlear tube and the membrane within it gives it very specific physical properties. Sounds conducted into the cochlea cause the basilar membrane to vibrate, but the location where those vibrations are greatest is exquisitely dependent on the sound frequency. High-pitched noises, like the song of Bill's meadow pipits, vibrate the basilar membrane most intensely when they are very close to the entry point of sound into the cochlea. The low-pitched tones of rumbling traffic affect the basilar membrane at the point furthest from the sound's point of entry. The membrane is ultimately acting as a frequency analyser, with different points of vibration matching to the frequencies heard.

But the membrane itself does not communicate information to the brain. Those vibrations still represent mechanical rather than electrical energy, and the work of transformation must still be done. Attached to the membrane is a fragile structure called the Organ of Corti, within which sit the even more delicate hair cells. Tiny fine strands, or hairs, only 20 micrometres in length, project like fingers from these cells, and when the basilar membrane vibrates, it is the deflection of these hairs that enables them to identify both the amplitude and rhythm of that vibration, precipitating an electric current, the start of the electrical journey to the brain.

So, we can start to see how the structure of the ear addresses some of the challenges involved in hearing when it comes to the analysis of frequencies and the conversion of mechanical into electrical energy. But what about the issue of volume? How is it that we can hear such a range of sound energies, from the drop of a pin to a roaring jet engine? There is one mechanism that both provides a degree of protection from very high sound energies and also addresses another issue, that of sound crossing from air to liquid. That mechanism is the middle ear, a seemingly unnecessarily complicated series of three bones that bridge the gap between the eardrum and the oval window, the entry point of sound to the cochlea. Of these, the stapes, a stirrup-shaped structure, is the smallest bone in the body, roughly 3mm in diameter and weighing in at 6 milligrams. In combination, these three bones act like a hydraulic system, overcoming the issue of sound transmission between air and fluid; sound pressure from the relatively large eardrum is focused by these structures onto the much smaller oval window. The bones themselves have some influence on our hearing. As light as they are, they do have some inertia, which prevents them from efficiently transmitting high frequencies, thus influencing our range of hearing. In fact, species sensitive to very high frequencies, like bats, have even tinier middle-ear bones to facilitate transmission of these ultrasonic energies. The exceptions to this association between middle ear bone size and hearing high-pitched noises are aquatic mammals, like dolphins, who famously generate high-frequency clicks and other sounds to echo-locate and communicate. They, of course, do not need to worry about the problem of getting sound from air to liquid, and actually transmit sounds from the sea that bathes them into the inner ear through their lower jaws.

The complexity of this air–liquid bridge, these three bones rather than one, permits some degree of control over incoming sound transmission. A tiny 6mm-long muscle, called the stapedius, attaches to the stapes and acts like the dampener on a car's suspension. When contracting, it limits the movement of the stapes against the cochlea, dampening loud noises and preventing damage to the fragile structures of the inner ear. We are not aware of the action of this muscle, although occasionally we see the impact of its failure. One of the more common neurological conditions is Bell's palsy, damage to the facial nerve resulting in weakness of one half of the face. But within the facial nerve are carried a few nerve fibres that supply this miniscule muscle, and when the nerve is damaged, people affected often note that loud sounds seem even louder, occasionally unbearable. The stapedius muscle is not under conscious control, however, and is contracted by a reflex in response to very loud sounds, or when we are talking. But this reflex cannot keep up with crashes, gunshots or other loud noises out of the blue; repeated exposure to these sudden explosions of sound energy are difficult for the ear to defend itself from, so they are much more likely to cause damage.

In addition to the stapedius, the ear has an alternative method for dealing with the massive range of volumes it encounters. Some of the hair cells perched on the basilar membrane have the ability to amplify sounds. These hair cells, called outer hair cells, possess an extraordinary property. Their outer linings – their cell membranes – contain a protein called prestin. Prestin acts like a subcellular motor, causing the hairs on the cells to move. When stimulated by sounds, the prestin causes the hairs to intensify their movement, exaggerating their deflections and amplifying the ability of these cells to detect even tiny amounts of energy – the rustle of a leaf, an exhalation of breath. But this super-sense comes at a cost. These outer hair cells are easily damaged, and their injury can result in severe hearing loss. Despite the protection that the stapedius affords, repeated sudden loud noises cannot be compensated for, hence the hearing loss that afflicts so many of the musicians that Bill knows.

Despite the drum kit taking pride of place in the room that Bill and I sit in, his hearing loss is not likely to be a direct result of noise damage alone. Now seventy-eight years old, he suffers the hearing loss that affects so many of us as we get older. In fact, presbyacusis – age-related hearing loss – is so common that it is almost a guaranteed aspect of ageing. By some estimates, 100 per cent of people over the age of eighty will have significant hearing loss. Even beyond the age of fifty, some 40 per cent suffer from it. Twice as common as cardiovascular disease and five times more common than diabetes, hearing loss constitutes one of the most familiar problems of ageing. Many factors contribute to this diminishment of hearing. The outer hair cells that are so vulnerable to noise-related damage are lost with age, making the combination of age and exposure to loud noises a particularly unhealthy one. Genetics, ear disease, and exposure to drugs that damage hearing all contribute too.

The nature of the hearing loss is especially cruel. Presbyacusis leads to the loss of high frequencies in particular, resulting in increasing difficulty in discriminating speech, or, in Bill's case, the high-pitched songs of his beloved birds. This loss of higher frequencies makes hearing consonants especially problematic, transforming clear speech into mumbling. Its impact on speech discrimination can have devastating consequences, with increased listening effort and hampered communication driving avoidance of social interaction and exacerbating loneliness. This process is caused not just by damage to the hair cells in the cochlea. As we age, all sections of the auditory system suffer. The structure of the cochlea itself shows evidence of ageing, as do the nerve fibres that transmit the electrical impulses from ear to brain. Even the areas of the brain responsible for perception of sound and speech show changes, both in terms of structure and function. The brain's ability to focus attention on the elements of noise we want to hear, and block out those aspects we do not, also deteriorates, the process of picking out a conversation in a noisy room being reliant not only on the ear but on the brain too.

Whether these changes in the auditory cortex are as a consequence of reduced inputs from the ear, or part of the ageing process simultaneously affecting the ear and brain, remains to be clarified. But what has become clear in recent years is the association between age-related hearing loss and more widespread brain changes, in the form of cognitive impairment, and it is not necessarily just the auditory cortex that is affected. Large studies have shown that this hearing loss is associated with dementia, and the worse the hearing loss, the higher the risk of cognitive decline. That is not to say that hearing loss causes Alzheimer's disease, nor that if your hearing is poor, you will get dementia. It is entirely possible that dementia and hearing loss go hand in hand due to a common underlying cause, or, as another hypothesis proposes, that the efforts of trying to hear with imperfect ears diverts brain resources from other areas of cognition. But one particular theory is gaining traction in the scientific world: that sensory deprivation not only causes reallocation of cognitive resources, but also drives decreased socialisation, reduced communication, and depression, all of which contribute to cognitive decline. That wonderful telepathy – '"beaming your thoughts into the other person's head" using as your medium a form of "invisible vibrations"' – is at the core of our connection to our partners, our families, our colleagues, our society. Connecting with others is at the heart of what makes us human. When hearing is lost – that ability to express or hear our inner thoughts, feelings, desires and viewpoints – it also results in the loss of place, loss of connection, and maybe ultimately the loss of self.

1996, autumn. As I strode alongside my consultant through the walkways of the Stonebridge Estate in north-west London, I was beginning to regret my choice of clothing. I was wearing the uniform of the male medical student – shirt, tie and chinos, armour in the constant battle that medical students fight in order to assume a degree of gravitas, clawing desperately for a semblance of professionalism on the wards. A jacket or blazer would be perceived as too much, presumptuous, and would be reserved for junior doctors about to become consultants. Here, among the decaying 1960s high-rises in a housing estate with a reputation for crack cocaine, gang violence and shootings,

I realised my mistake – my choice of tie, in particular. A gift from someone, the tie's design featured a bookcase of leather-bound books, and what I had previously considered a learned look now felt like impending victimhood. My psychiatrist boss had taken a different approach, his dark suit hidden under a rumpled mac. His absence of a tie, I later learned, was in response to having been strangled by a couple of patients in the past. As we stepped into a cramped steel coffin of a lift, smelling strongly of urine, I felt a wave of relief that we were no longer quite so exposed. Somewhere several floors above, we came out into an open passageway and made our way to one of the flats, identical but for the front doors, in different colours and varying degrees of disrepair. My boss knocked on the peeling door, and after what felt like an age, I heard a chain being drawn and the door being unlocked.

It was my first week attached to psychiatry, and the previous few days had been confined to the lecture theatre. We had learned about the various aspects of taking a psychiatric history, how to assess for risk of suicide, what constitutes psychosis and neurosis. But to date, I had yet to meet a patient. My consultant, a community psychiatrist, had brought me along to undertake a home visit to one of his patients, a woman in her forties with chronic schizophrenia. As the door opened, I could see a dark corridor – and a face. The woman peered out at us, saw her doctor, barely glanced at me, and turned round, leaving the door open behind her. I could immediately see the impact of her treatment. Her gait was slow, with limited swinging of the arms, and every movement was restricted, at a snail's pace. When we eventually sat in her living room, which smelled strongly of cigarette smoke with damp mould undertones, I could see her blinking sluggishly and infrequently, a paucity of facial expression. Even her speech, as she greeted her consultant but deliberately ignored me, was like a 45rpm single played at 33rpm speed. All this, the mark of years of antipsychotic treatment, producing Parkinson's disease-like side effects.

She sat in a worn armchair, the armrests blackened by years of dirt, the detritus of a difficult and chaotic life around her. On the floor, next to her, was an overflowing ashtray, the carpet around it pockmarked with burns from misplaced cigarette butts. My consultant sat on a small sofa immediately opposite her. Once again I was envious of his old coat, shielding his clothes from the unspecified stains of varying colours decorating the furniture like a Jackson Pollock painting. I perched delicately on the other end of the sofa, trying to avoid too much contact with the seating and any eye contact with the woman. I sat in silence, nodding briefly as I was introduced to her.

The consultation started, my consultant gently enquiring as to how the woman was. With each question, a small delay, and then a brief and slow response. She spoke quietly and hesitantly, and I strained to hear her words. I relaxed a little as I realised that this was not so different from other consultations I had sat in on. Lulled by the gentle exchange between them both, my mind began to drift as I looked around the flat – but my attention was suddenly brought back into focus when the woman let out a dramatic shriek, then a shout of 'Medical student, don't do that!' I could feel my heart racing, and colour rising to my cheeks, but when I looked at the psychiatrist, he had not flinched or given any outward indication that she had said anything untoward. Their conversation continued as before, in hushed tones, no evidence that it had been punctuated by her outburst. But a few moments later, again, a shout of 'No, no, no, medical student! Don't do that!' And as the minutes passed, her outbursts increased in frequency: 'Don't say that, medical student!', 'No, I won't, medical student!', 'Why are you saying that, medical student!' As I sat there, silently observing, she said she could hear me encouraging her to harm herself, stab herself, hang herself. Needless to say, the visit did not last much longer. I still hear the accusatory, hurt tone in her voice now – 'Medical student!'

The bookshelf tie died that day, but my fascination (combined with dread) for psychiatry was born: the nature of the human mind, the breadth of human experiences, but also the horror of psychiatric conditions and the fine line between sanity and 'insanity'. It is perhaps the evidence of an immature mind, but throughout my psychiatry placement I could not shake the feeling that there is blurring between the margins of normality and psychiatric illness, a deeply unsettling thought.

That incident ('Medical student!') was my first encounter with auditory hallucinations – the woman hearing my voice telling her to kill herself. Of course, everyone will be familiar with the concept of imaginary voices – hallucinations of speech as a manifestation of mental illness, just like the visual hallucinations of psychosis discussed in the previous chapter. Before the advent of mobile phones and earphones, the sole man wandering down the street in loud conversation with himself was a marker of psychiatric disease. But auditory hallucinations are not always a manifestation of mental illness. Indeed, almost all of us will have had auditory hallucinations, albeit of a less dramatic nature. We hallucinate every night, in dreams of speech or music; more obviously, the ringing in the ears after a few hours at a nightclub or concert, or the buzzing accompanying an ear infection. These both meet the definition of a hallucination: the apparent perception of something not present. More specifically, these latter experiences represent tinnitus, derived from the Latin verb *tinnire* (to ring) – a hallucination, the perception of a ringing, buzzing or hissing, in the absence of a real sound in the outside world. Even beyond the temporary phenomenon induced by a wild night out, tinnitus is incredibly common. It is estimated to affect some 10–15 per cent of adults, more so with increasing age. This correlation with older age is not surprising, given the strong association between ageing and hearing loss, but hearing loss is not a prerequisite for the condition. Some people with severe tinnitus may have normal hearing, while many people with profound hearing loss do not experience tinnitus. Other factors that correlate with tinnitus include damage from persistent noise exposure or drugs, and disorders that may influence nerve function.

But why should damage to the hearing apparatus, be it the middle ear, the cochlea or the vestibulocochlear nerve – the nerve that carries the electrical impulses from the ear to the brainstem – cause a gain of hearing, as well as a loss? Well, we have already seen one mechanism that amplifies very quiet sounds – those outer hair cells in the cochlea, with their molecular motors. But these are not the only way the auditory system amplifies sounds. It appears that there are mechanisms in the brain, too, that become active in the face of quiet or silence, increasing the sensitivity of auditory cells in the auditory cortex to such an extent that they perceive sound that is not actually there. It is this increased sensitivity of the auditory cortex that is thought to underlie tinnitus. Rewiring of the auditory centres of the brain, so-called 'plasticity', may be the true basis of chronic tinnitus. So, while the initial origin of tinnitus may be damage to the inner ear or nerve, tinnitus itself is a product of the brain; if you cut the vestibulocochlear nerve of someone who has tinnitus, their auditory hallucination will persist, even without any inputs at all from the ear. There are clear parallels here with Nina and her visual hallucinations of colours and shapes following her loss of vision. The brain wants to be able to hear as much as it wants to be able to see, and when deprived of visual or sound inputs, it will create its own visual or auditory world.

Bill Oddie's loss of hearing has also been balanced by a gain, as with Nina's vision, but Bill's compensation is not tinnitus. His condition is rather more unusual. It started rather suddenly, about two or three years ago. 'I was in the house, as we are now,' he says, 'and I just thought, *Oh, somebody's playing music very loudly next door.* It sounded like someone playing a record or radio. So I went over to the wall, just curious to see what they're playing. And I thought, *That's strange. It's not quite the same [music] here.* I moved round different parts of the house, and discovered similar music going on.' For weeks, Bill continued to move to different rooms and corners of the house, trying to pinpoint the source of the music. 'I kept going up to my wife, saying, "Did you put the radio on at four in the morning?" And she'd say, "Of course I didn't,"' he chuckles. 'I think my wife got fed up with me saying, "Did you hear somebody playing radio last night?"' I ask him if he recognised it as a hallucination fairly quickly, but he tells me he thinks he was initially in denial.

That was, however, just the start of the musical accompaniment to Bill's life. Since then, he has acquired a soundtrack that follows him almost everywhere. It's more noticeable with silence and solitude, but often present in the background the rest of the time too. I ask him what kind of music it is. Bill reports it as being a rather particular sound: a hint of brass-band-type music. He says there's nearly always a lead trumpet playing a high note – 'which happens to be one of the sounds that I really dislike.' But while he describes a consistency to the presence and tone of the music, there is an indistinct flavour to what he is hearing, and he thinks that his perception of what the music represents has shifted with time. Sometimes he questions whether he is actually able to make out a clear tune.

Over the past year or two, he has rarely experienced purely instrumental music. 'There are almost always singers, always a male voice, or a relatively small male-voice choir; very occasionally a female voice, or another voice that I'm picking up, like an announcer. It sounds like someone playing a radio, but it's an old-fashioned style.' And as time has passed, he has also begun to recognise the repertoire – 'sort of rollicking. Wonderful country songs from after the First World War.' All of a sudden, Bill bursts into song, a few bars of 'All the Nice Girls Love a Sailor' in his own distinctive voice, and once again I feel like I am listening to a Radio 4 comedy panel show. He goes on to describe a jukebox selection of songs – 'Only, I can't choose them!' – singing snatches of 'Daisy, Daisy', 'Blaydon Races', 'She'll Be Coming Round the Mountain', 'Rule Britannia' and the national anthem. 'I was hoping to recognise something more modern. Something more up to date. I suppose that disappoints me. It doesn't make any sense to me – why the music is from a particular era.' He would prefer an internal Alexa – 'one of those things you shout at' – and would opt for jazz, jazz-rock or Americana.

I ask Bill if the musical hallucinations are getting worse or if they ever stop him doing anything. 'I don't think they are getting worse,' he says, 'but I think I'm getting more annoyed with them. If I'm up in my office and I hear it, it can go on a bit. Once or twice, I have been known to say "Shut up!"' he laughs. 'Generally, though, it is not affecting my life, really.' I wonder if it ever completely goes away. 'Oh, it's not there now. It was here half an hour ago, before you came. It knew you were coming,' he snorts. '"Lads, tea break's over!" I haven't lost a certain sense of humour about it. I would if it was driving me nuts.'

I am hesitant to ask the next question. Bill has been on record over the years talking about his struggles with depression and bipolar disorder in his life, and I ask him if he thought he might be going mad when the hallucinations started. 'No,' he chuckles. 'I've already been mad! I'd already had one or two genuine supposedly bipolar experiences. But this was different. I wasn't particularly depressed. I was just, "Where the hell is that coming from?" It was straight out of the blue.'

When he realised what the music was, he went to the internet and ordered some books. 'I found this one by some American guy . . .' I tell him that the book, *Hallucinations*, was by a north London boy who moved to the US. Oliver Sacks grew up a couple of miles away, in Mapesbury Road, Kilburn. As a child, Sacks would swim with his father in Hampstead swimming ponds, a stone's throw from Bill's house. 'Yes, that's him! What did strike me was that there seemed to be not much consistency about things – you know, that some people heard choruses and choirs, some people heard some other type of music, some people just heard horrible noises and so on.'

Bill's soundtrack is a hallucination, and it is intrinsically different from the hallucinations of psychiatric disease, like those of the patient I visited as a medical student, sitting in her armchair, shouting 'Medical student! Don't do that!' While the hallucinations sound real to Bill, like music playing next door, he knows they are hallucinations. His grasp on reality remains firm. The singing and trumpet playing is more akin to tinnitus; he has the perception of sound in the absence of an external stimulus, but rather than hearing a ringing, buzzing or whooshing, it is more complex, more nuanced, more melodic. While thought to be rare, this is perhaps an underestimate: a recent study of people attending a hearing clinic found that over 5 per cent of patients reported hearing music. Perhaps people are disinclined to report these symptoms for fear of being thought 'crazy'. Bill's musical experiences are really rather typical, with people describing big-band music, church choral music, trumpets or bugles, occasionally even elevator or country music. As time passes, however, many patients report that the music tends to fragment, with shorter and shorter melodies being heard. And while musical hallucinations are seen in individuals with psychiatric disorders or with neuro- logical diseases such as dementia, seizures, brain infections or tumours, it is also seen in entirely healthy individuals. The most common factor is hearing impairment, although even that isn't always the case.

Not surprisingly, when it comes to those patients with neurological disorders, many of these relate to damage to the auditory centres of the brain. Indeed, there are specific areas of the brain that seem to encode music. Wilder Penfield was an American–Canadian neurosurgeon who would electrically stimulate the brains of awake patients before removing areas of the brain substance, to understand if the regions he was about to remove served an important purpose. Through stim- ulating either the left or right superior temporal lobe, Penfield induced musical hallucinations in eleven of his patients, elicit- ing Christmas carols, radio-programme theme tunes, a piano and an orchestra.

Once again, there are very clear parallels with Nina and her Charles Bonnet syndrome – visual hallucinations in the context of eye disease. Tinnitus is the equivalent of the simple shapes or colours Nina saw, while Bill's jaunty tunes from between the wars correspond to more complex visual phenomena akin to Nina's zombie faces or the classic Lilliputian figures described by Charles Bonnet himself. Presumably, the nature of auditory hallucinations – whether they're simple or complex – depends on the specific areas of the auditory regions of the brain that show increased activity, usually in response to changes within the ear itself.

As with vision, there is a unifying theory as to how these sound phenomena may arise – musical hallucinations, tinnitus and the hallucinations associated with psychosis ('Medical student! Don't do that!'). This model of the brain is one that bridges the gap between neurology and psychiatry, providing a plausible biological explanation for mental illness. The theory is one we have touched on before – that the brain is ultimately a prediction machine, taking inputs from its environment but also sending outputs down to its information-gathering apparatus. Our brains do not have the capacity to recreate our environment from scratch at every single moment of the day. Instead, we predict the most plausible explanation for what we perceive to be happening, based on an internal model of our world as we understand it. When balanced, streams of information from the outside in (raw sensory information) and from the inside out (our expectations) allow our senses to work perfectly. Occasionally, of course, something unexpected will happen. Our predictions, our expectations, will be incorrect, and we will learn from these moments. But when the system is out of kilter, when either the inputs are too limited, as in hearing loss, or when the outputs of our brain – those predic- tive models of the world – are too intense, as in psychosis, this results in hallucinations or delusions. If your brain is so certain of its predictions that it ignores sensory information coming in, it may hold false beliefs or experience false perceptions.

Those very mechanisms that are designed to allow us to better understand our world ultimately end up confusing our sense of reality rather than clarifying it.

In fact, this view of how our brain works, and how we perceive, is having some surprising outcomes. Drugs like ketamine, psilocybin (the active ingredient of magic mushrooms) and LSD have long been used and abused due to their mind-altering consequences. The changes to perception that 'trips' on these drugs entail can also be understood in terms of alterations of prediction. Consider someone on a 'trip' staring at a house plant, the fronds of green slowly morphing into writhing snakes, then rivers of green, then back to a house plant. It seems that these drugs exert their effects through removing the predictive constraints of our daily life, leaving us free to consider other possibilities, other interpretations of our sensory data. In essence, what these drugs are thought to do is to alter that balance between inputs and predictions, promoting chemical changes in the brain that free us up to perceive in a less rigid way. This model of our minds introduces possibilities for the use of these kinds of drugs in the treatment of mental illness. These drugs create the opportunity for us to perceive situations in different ways. For example, your anxiety may be rooted in rigidly predicting what will happen tomorrow, next week or next year. Your depression may have a basis in overly rigid predictions about your own self. Allowing you to consider different predictions and different realities may provide the opportunity for change. Indeed, evidence is now emerging to support the use of these drugs in the treatment of depression, and a ketamine-based nasal spray has been brought to market for treatment-resistant depression.

Once again, we are left with the knowledge that, as with our other senses, reality and perception are two very different things.

The mechanical energy passing through the air that is sound is something tangible, measurable, real. Its relationship with what we perceive, what we actually hear, is much more complex. As we have seen, that relationship depends upon our ears, upon our brains, and can be fundamentally altered by damage or disease. It is even modulated by normal ageing. And our expectation of our world, our prediction of what we should be hearing, is occasionally a stronger determinant of what we hear than sound itself, as with Bill Oddie's musical hallucinations, the tinnitus that we may hear or even the experience of being sure we have heard the doorbell or the mobile phone ringing when we are waiting for a delivery or an important call. And if this all sounds somewhat implausible, running counter to your view, I urge you to view a simple video that is freely available on the internet. Search for 'the McGurk effect'. You will be met by a video of a person repeating 'Bah, bah, bah' over and over. At a certain point, the person will begin to say 'Fah, fah, fah', the movements of their lips changing in synchrony as they enunciate a different consonant. But close your eyes and immediately you will realise than the soundtrack is still 'Bah, bah, bah'. Open your eyes again, and you will again hear 'Fah, fah, fah' as you watch the person's lips move. What is playing from your laptop's speakers is completely unchanged throughout; only the video changes. The expectation of what we hear, suggested by our vision, directly influences what we actually hear.

No matter how many times I watch this video, I am always just a little shocked; surprised that my own senses are so suggestible, so flawed and fallible (and, in truth, they're likely to become more so as I get older). To hear it with my own ears is usually to be convinced of reality, but perhaps we all have a misplaced faith in our own experiences.

*

Postscript: I contacted Bill to ensure all was correct in this chapter. His email starts in his familiar style: 'D ear – accidental but appropriate mistype – Guy'. He tells me of his unpleasant experiences of 2020 and the pandemic, then continues:

A final comment may be in order. So . . . Well, the male-voice choir are still going strong. The lyrics are less clear and the melodies less familiar, but the overall style is consistent. As well as the vocals, I get 'riffs' played by what sounds like a kazoo, or even a 'Stylophone', as pioneered by Rolf Harris. Years ago! Surely not! I remain both surprised and narked that what I hear is NEVER like music I have listened to or enjoyed. Who or what is selecting the repertoire?! I need a new cranial DJ please. Oh no, NOT 'Land of Hope and Glory' again.

Is there a cure?

5

IN THE KINGDOM OF THE BLIND

'I don't think it had ever before occurred to me
that man's supremacy is not primarily due to his
brain, as most of the books would have one think.
It is due to the brain's capacity to make use of the
information conveyed to it by a narrow band of
visible light rays. His civilization, all that he had
achieved or might achieve, hung upon his ability to
perceive that range of vibrations from red to violet.
Without that, he was lost.'

John Wyndham, *The Day of the Triffids*

There are patients in one's medical career that stand out in
sharp relief from the constant never-ending stream of humanity – and its ailments – that flows through the doors of the
hospitals I work at. I would love to tell you that they are all
people for whom a heroic neurologist has snatched victory from
the jaws of diagnostic defeat. Where the (usually bespectacled,
and slightly nerdy) protagonist strides onto the ward clutching
his trusted tendon hammer, briefly casting his eye over the
despondent patient languishing in bed, before pulling out a rare
diagnosis in a jaw-dropping display of logic, intellect and expertise. There is a quick course of treatment, then resurrection,
a Lazarus-like recovery resulting in a previously bed-ridden
invalid striding out of the ward.

Unfortunately, this type of story occurs very infrequently. But there are countless others who stick in my mind for other reasons – disasters, demise, horror. The young Japanese girl, a tourist in London, admitted with acute abdominal pain on my first-ever night shift, whose belly I saw rapidly expand as she bled out into her abdominal cavity from a rare tumour. The man in his sixties I was called to see in the middle of the night, as a general medical registrar, who was admitted with chest pain thought to be cardiac in origin. He proceeded to vomit litres and litres of bright-red fresh blood, necessitating transfusion of massive amounts of fluids, plasma and red blood cells, with me standing by the bed squeezing the bags of blood to push it into his veins more quickly, all the while dodging the jets of scarlet projectile vomit. The young woman brought into A&E in cardiac arrest, having been chatting to her hairdresser one moment as the hair dye was applied, then suddenly going into anaphylactic shock from the dye – an explosive allergic reaction. She expired on the resuscitation trolley, the red hair-colour staining the white linen beneath her head like a smear of fresh blood.

Sometimes they stick in your mind because you can perceive something of yourself in them, perhaps by being of a similar age, having the same interests or because you can imagine being friends in the outside world. Occasionally they are memorable for other reasons, like the young woman I met who had suddenly developed blindness and paralysis of the legs, whose lumbar puncture I was tasked with – the first of my career. Fifteen years later, as a consultant, I met her again on the wards after another admission, her face largely unchanged, although both of us were slightly greyer. On looking through her medical notes, I found my own scrawl documenting the procedure from over a decade earlier, and recalled the wave of euphoria at my first successful spinal tap. When I reminded her of this, she told me she remembered my voice.

Other patients stick in the mind because I have cared for them for a long time, witnessing their unstoppable decline and the impotence of modern medicine in treating brain tumours, motor neurone disease, horrible genetic conditions; people like Rahel, in Chapter 1, or Dennis, one of the first patients I looked after as a junior doctor. He had been admitted onto our urology ward in urinary retention on day three of my life as a doctor. Where he had come from was a bit of a mystery, but Dennis was in the latter stages of dementia. He had been a merchant seaman in a previous life, and with his grey, bushy nautical beard kept neatly trimmed by the nurses, and his anchor tattoos, he had an air of Captain Birdseye about him. Twenty-two years later, I can still see Dennis' face in my mind's eye, as if it were yesterday. He lived in the moment, with little recollection of the passage of time. Every morning I would go into his room to be greeted by a manic toothless grin and a shout of 'Good morning!' at the top of his lungs. Always cheery, for no good reason, with a gentleness of face and demeanour. His urinary retention had long been sorted out, and my bosses were eager to get him off the surgical ward so that they could admit other patients for surgery. In the words of Samuel Shem's novel *House of God,* a dark and hilarious satire that almost every doctor is familiar with, he was a true GOMER (Get Out of My Emergency Room) – the first rule of the *House of God* was 'GOMERs don't die', and apart from his urinary symptoms and dementia, Dennis was in fine fettle. But he had nowhere to go to, and my mission was to find him a destination. For three months we were locked into bureaucratic and administrative purgatory, with endless cycles of finding a home for him before being told that the promised bed in a nursing home was unsuitable or unavailable. All the while, I was greeted every day by a hollered 'Good morning!', a perpetual Groundhog Day.

But despite Dennis' limited understanding of the passage of time, his unchanging clinical environment began to take its toll and he started to show little flashes of frustration, undertones of anger in his morning greeting. In the last week of my post, we finally heard the news that Dennis would be discharged. A bed in an appropriate nursing home had been found and, importantly, funding had been agreed. Dennis' sentence on the urology ward was due to come to an end, the release date set for the coming Tuesday. We were all delighted, my mission – for Dennis to leave urology before I did – looking hopeful. Even Dennis seemed a little more upbeat, perhaps sensing the lift in the atmosphere around him. And so, as I walked out of the hospital on Friday night for the weekend, knowing that my post was to finish the following Tuesday, with Dennis having departed to a home on the very same day, I had a slight spring in my step. Coming in early on Monday morning, I anticipated my penultimate 'Good morning!' as I started the ward round. But when I got to Dennis' room, the bed was empty, stripped of sheets, all personal possessions gone. Initially I assumed there had been a miscommunication, that he had departed to the nursing home earlier than agreed, a rare example of efficiency. But I soon learned that Dennis had departed to an alternative destination. Over the weekend, he had suddenly experienced a cardiac arrest, and died. A prisoner of circumstance for the last three months of his life, within touching distance of the finishing line but falling at the last step. He had broken the first rule of the *House of God*.

There are two other categories of patients that stand out. The first is those for whom I could have made a difference but wasn't able to, where I question whether, had I done something differently, the outcome would have changed. In my own specialties, of epilepsy and sleep, there have been patients found dead at home – the sudden, unexplained death of epilepsy or the more readily explained death from drowning due to a seizure in the bath; patients with sleep disturbance committing suicide, tortured by their own minds; occasionally brain tumours, small and continuously monitored, suddenly and unexpectedly transforming to something much more malignant and aggressive. I am not eager to dwell on these cases in the pages of this book, but suffice to say that decisions I have made – the choice of starting or not starting a drug, of scanning someone now or in three months' time – still come back to haunt me in the small hours months or even years later, names and faces seared into my memory.

But the other category of patients are those affected when medics have inflicted damage – not through their deeds, but through their words; what I term 'iatrogenic communication' or *communicationis iatrogenica imperfecta*. Nowadays in medical school, there is a huge focus on communication skills, the ability to extract and convey information in a way that is efficient, kind and intelligible, but it was not always this way. Even as relatively recently as my own time at medical school, we would roll our eyes at this, eager to get to the nitty-gritty of blood, surgery, fancy drugs, expensive tests. But I now understand that one can be the most academic doctor, reading avidly and able to quote the latest studies, knowing every single cause of a certain positive blood test or being highly skilled with a blade, but that in itself does not make one a *good* doctor. Without the ability to talk on a human level, to empathise, understand, grieve alongside your patient, to connect on any level, even superficially, it is impossible to doctor well. A major part of our role is to convey risk, uncertainty and judgement. To do this while understanding that the person sitting in front of us is more than just a patient or disease, that outside the confines of the consulting room's four walls they are a parent, a spouse, a colleague, with competing responsibilities and psychological pulls, all of which influence their health and the decisions they make about it. Without empathy and communication, we would be no better than robots. But over the years, I have seen some truly awful examples of unthinking communication, where the words of a doctor have caused as much harm as the inexpert wielding of a scalpel or the prescription of the wrong drug. There's the consultant surgeon whipping back the bed covers to expose a patient's blood-starved leg, bluntly saying, 'It's got to come off!', before striding away to the next bed. Or the registrar on the elderly care wards whose ability to communicate with empathy was so awful that we juniors used to have to visit all the patients after the ward round, trying to repair some of the devastation he had wreaked. In particular, I recall one poorly elderly lady who had been admitted with chest pain. She was upset not only by her predicament but also by the fact that her husband had died on the very same ward some six months earlier. She was desperate to get out of there. The Butcher – nicknamed for his cackhandedness at empathy and kindness as much as for his clinical skills – simply nodded, listening as her tears fell, before smiling and attempting to reassure her with, 'It's okay! People die in hospital.' He then turned round and walked on, satisfied that he had made her feel better, the sound of escalating wailing ringing in my ears as I followed him to the next bed. To this day, I still question how he thought that his response was in any way helpful, although, in his defence, I think his interactions with patients were clearly borne of ineptitude rather than malice.

As a neurology registrar, for a period of time I would

regularly take the Tube on a Friday afternoon from St Thomas'
Hospital to Moorfields Eye Hospital. There, among the vast
numbers of individuals with glaucoma, cataracts or retinal
problems, would be patients for whom the ophthalmologists
could do nothing, either because the problem with their vision
was outside the eye, or because, ophthalmology being primarily
(but not exclusively) a surgical speciality, the problem could not
be fixed with an operation. So, my boss at the time, a neurologist by training but particularly interested in vision, would
travel between various hospitals in London, providing a service
for the evaluation, diagnosis and management of these complex
patients. The Friday p.m. clinics at Moorfields would be a combination of service and training, in a large room equipped with
various bits of kit and multiple desks, where the neurology and
medical ophthalmology trainees would see patients referred in
from the rest of the hospital. My consultant would circulate,
we would present the patient and our conclusions to him, and
wait for the nod or a shake of the head. There was a degree of
theatre to this too. He was world-renowned and people would
travel from all over to observe his clinics. For each patient, he
would hold forth in front of the assorted doctors and patients,
his brain an encyclopaedia of clinical information. He would
draw our attention to a tiny clinical detail, or describe some
study he had done twenty years earlier that threw light on the
condition at hand; a veritable walking textbook. The majority
of the patients we saw had visual migraine, stroke or conditions
affecting the muscles or nerves to the eye, but there were a fair
few rarities thrown in.

During one of these clinics, I was asked to see a gentleman
in his early sixties, accompanied by his wife. As I sat down
with him in a corner of the huge communal clinic room, the
lack of privacy astonishing, he began to tell me how his vision

had been deteriorating for the past two or three years. He was finding it progressively difficult to read, and just felt that there was something else wrong, although he could not precisely put his finger on it. Over the course of the preceding two years, he had been back and forth between high-street opticians and the ophthalmologists. All tests had returned normal, although some corneal abnormalities had been identified. Despite several procedures to his eyes, his vision had not improved. If anything, it was worsening, which was why he had been referred to this clinic. When I examined him, it was clear that his near and far vision were entirely normal, as were his eyes. But when I asked him to read some text, he found it almost impossible. Becoming suspicious that this was indicative of a brain rather than eye problem, I pulled out a small green book, a tool we use for cognitive testing. Among its pages are pictures of faces to test facial recognition, line drawings of various animals and tools to recognise visual meanings, like pictures of boys playing with sandcastles on a beach. And as we worked our way through these tests, it became clear that while the gentleman's eyes were normal, the visual cortex – the part of the brain drawing meaning from visual inputs – was far from working normally. He could not make out broken letters and had difficulties recognising simple visual objects – a teapot, a shoe, a padlock. Putting together the time course of the progression of his complaints, a previously normal CT scan of the head, and the nature of his cognitive problems, I was fairly convinced that this was an unusual presentation of Alzheimer's disease, a diagnosis subsequently proven correct. I asked him what he thought might be going on, but it was obvious that neither he nor his wife had any inkling of this – they were convinced it was some as-yet unrecognised eye complaint.

The time came for me to present my findings to my boss, and

as I began, other doctors, ophthalmologists, gathered round. Out of the corner of my eye I could see the patient and his wife listening carefully, hanging on every word; I have always been uneasy with this aspect of clinical life, as it is sometimes deeply unhelpful to discuss a patient's case in front of them. I got to the end of the findings and paused, fearful of outlining my conclusion, knowing that the man and his wife did not expect the diagnosis I had made. After a few seconds, I was asked what I thought was going on, and I responded, 'I think he has beta-amyloid deposition,' using a highly technical phrasing implying Alzheimer's disease but allowing me to shield the poor couple from having the diagnosis broken in such a public and brutal manner. There was another brief pause, before one of the ophthalmologists blurted out, 'What? You think he has Alzheimer's disease?' The look of horror and dismay on that man's and his wife's faces has never left me, the cruelty of that moment carved deep. Every time I think about that awfulness, I feel a rising tide of nausea.

The case clearly illustrates the damage we can do, the addition of insult to the original injury of the disease. But it also exemplifies something else: that, as with visual hallucinations, such as Nina's, or other visual phenomena, the loss or diminishment of vision may also be due to problems outside the eye. Eye disease, as all ophthalmologists will recognise (although may occasionally miss), is not the sole cause of loss of vision. In our speciality bunkers, we all have a tendency to see a symptom through the prism of that speciality. While the eyes capture light, and turn that light into electrical signals, it is in the brain that we 'see' – where sense and meaning is made of the outside world. At each level in the nervous system, the degree of complexity increases, the richness and consequence of our visual universe added in layer upon layer. And the way the problem or defect manifests depends on where in the nervous system it originates. In the case of the man just described, a gradual, diffuse deterioration of the visual areas of the cerebral cortex, where Alzheimer's disease rarely has a predilection, results in an inability to interpret the complexities of writing or the significance of visual objects. But in other people, damage to different areas of the visual system may manifest very differently.

I have been seeing Dawn now for many years. An unknown genetic mutation has caused her to develop multiple 'benign' tumours inside her head, tissue derived from the outer lining of the brain called the meninges. I use the term 'benign' in inverted commas as, although these tumours do not spread to other regions of the body, in Dawn's case they are steadily robbing her of her sight, as well as causing other complications. As they gradually expand, over the years they are relentlessly compressing the optic nerves, those thick cables conveying visual information from the retina to the brain.

Dawn's problems first came to light when she began to experience difficulties in the classroom. She was working as a teaching assistant, and noticed that she was having difficulty reading. It all started very subtly. 'I was getting headaches and I wore glasses anyway, so I presumed I needed a new prescription,' she tells me. She ignored her worsening vision for a while, but recalls, 'On one occasion I was asked to read a notice to the class, but realised that I could not read it. I passed it to one of my colleagues, pretending I needed to go and do something else.' But it was only when she eventually went to an optician that she realised there was a problem. What Dawn initially neglects to mention is why she actually took action and booked herself an appointment. She eventually confesses, 'It got to the stage where my driving was becoming quite bad.' She pauses briefly. 'I did actually have an accident. But I blamed that on the sunshine.' She crashed into the back of a stationary bus. Despite Dawn telling herself that it was simply a case of being dazzled by the sun, her husband, Martin, insisted that she go to an optician, under threat of divorce. 'In a pretty forceful way,' Martin adds. The optician noted worrying signs at the back of her eyes: a swelling of the optic nerve as it exits the eyeball, indicative of raised pressure inside the skull. She was rapidly sent to a doctor. The initial view was that this represented a condition known as idiopathic intracranial hypertension, seen most frequently in young, often overweight women, in whom, for unclear reasons, fluid absorption from the brain is impaired, causing pressure to build up. 'The doctor I saw at the hospital actually said that they thought it was a condition that was recti-fiable. He said that they should be able to do something about it. So I suppose you walk away with a certain air of positivity . . .'

But a phone call from Dawn's GP dashed her hopes: it was another case of *communicationis iatrogenica imperfecta*. I ask her what the doctor told her. 'Basically, she said, "We've got your scan results back and we found five tumours." And that was pretty much it.' Dawn was devastated – she was alone at home when she received the news, no one to talk to, not knowing if these tumours were benign or malignant. 'A hundred and one questions floating around in your head at that point. With no answers.'

Try as we might to deny it, there is randomness in life. The course of our existence can pivot on a pinhead. The fragility of our lives, much outside our own control. We live in a world where we are told that if we don't smoke, don't overeat, don't drink too much and get plenty of exercise, we will live to a ripe old age, that we hold our own destiny in our hands. But as I think about people like Dawn in that moment in time, at the age of twenty-nine being delivered a telephonic blow shaking her existence to its core, I am reminded of the arbitrary nature of health and sickness.

There are personal reminders too. A few months ago, some expected news came – news that I was dreading. A friend of over twenty years had passed away from cancer – a neurosurgeon; serious, hardworking, operating on the brains of people with tumours. I had known him since our early twenties as very junior doctors – fun-loving, urbane, artistic, super-fit, devilishly handsome, filled to the brim with the vigour of life. At my wedding, he and his wife-to-be finally left the dancefloor at six in the morning, and at his own wedding we also danced through the night. One moment a successful surgeon, a father, a sportsman, the next moment with colon cancer, spreading to his liver and lungs. He was only a couple of years older than me, in his late forties when he died, leaving behind two young children and his wife.

In the last few months, he had been too filled with frustration and anger at the universe to see anyone, railing against the unfairness of it all. Emails and texts went unanswered. So I wrote him a letter, old-style, on pen and paper. I wrote of how his life was woven into our own, of his presence in my own and my wife's history. I wrote of his importance to all those people around him. But I also wrote of our views as doctors – that day in, day out, we see people whose lives have been blighted by illness, whose time on this earth has been cut short. But in our day-to-day practice, we protect ourselves. We believe it is us and them, that we will not be touched by the same hand of Fate that has anointed our patients. We stand apart, sheltered by an invisible shield, imaginary personal protective equipment.

And we rationalise this. As young doctors, we pretend that disease is simply a function of age. As we get older, and our patients get younger, this becomes increasingly difficult, but we persist in this self-deception; the reason for their illness is poverty, poor lifestyle or genetics, any angle that we can find to differentiate ourselves from them. While in our hearts we believe this, in our heads we know that this is not the full explanation. Of course, our lifestyle and personal circumstances influence our likelihood of death and disease – we all know that. But if we are honest with ourselves, we also are frequently witness to the randomness of life, the unpredictability, the throw of a dice. I think of Dawn, and the litany of people I have seen over the years, whose lives have been randomly and irreparably damaged or curtailed. As healthcare professionals, if we reflect, we know all of this, even if for the most part we ignore it. But for the majority of people, protected from this aspect of life unless it directly affects us, the arbitrary nature of health and illness goes unrecognised, especially if we are lucky enough to live in a country equipped with modern medicine, in control of our own destinies.

But there is something else that has shattered that illusion, for all of us. As I write, we are in the midst of the Covid-19 pandemic. In London, we are past the first peak, the hospitals slowly discharging patients, destined for home or, in too many cases, the morgue. We are living in a state of limbo, uncertain as to the future, whether Covid-19 will simply slowly peter out or if we will experience a second or third wave [the answer has since been forthcoming, and one wave has become an ocean]. This pandemic compels us to accept that there are forces beyond our control, that chance has a role. Our health, our vitality, can turn on the spin of a coin, a cough of a passer-by, a hand on a shopping trolley. We live with the knowledge that modern medicine is fallible, imperfect, and does not have all the answers.

That realisation – that we are mortal beings, fragile and vulnerable in ways that we never truly previously considered – has had a deep emotional impact on almost everyone. For healthcare workers on the front line, there is the expression of anxieties and fears not previously heard, a rising sense of doom at the prospect of their work. For many of us, there has been the nightly ritual of looking at the statistics, counting the deaths, trying to quantify them through the phrase 'underlying health issues', a euphemism for 'not like us'. Equally, there is the horror of seeing stories of death in people younger and fitter than ourselves.

For all of us, the pandemic has led to a re-evaluation of ourselves, the frailty of our bodies, in the recognition that the effects of this disease are unpredictable; a mild cough or loss of sense of smell, a prolonged stay in intensive care, or worse – its effects simply the luck of the draw. And whatever the outcome of this pandemic, this knowledge will persist beyond the virus itself. Much like the death of my friend, it is a brutal reminder that some aspects of life are simply beyond our sphere of influence, that life is intrinsically random.

And as I think of Dawn's wait for answers, as a mother of two young children, aged nine and six at the time, her life turned upside down by a phone call, I can only imagine her terror – the knowledge that growing inside your head are foreign invaders of your brain, robbing you of your sight, and possibly your life. In the course of every medical career, the times when you utter a sentence that you know is going to fundamentally change someone's life are all too common, and it is easy to become inured to these occasions. But the power of a single word – 'cancer', 'stroke', 'haemorrhage' – is almost unfathomable unless you are on the receiving end.

*

We all have blind spots, metaphorically but also literally. For every one of us, even with normal sight, the visual world is incomplete. As we look straight ahead of us, all of us have gaps in our vision. This is because the retinas at the back of our eyes – a semi-sphere of sensors coating the globe of the eye, capturing all the light entering through our pupils – are not entirely uniform. In the dead centre of retina, equating to the centre of our field of vision, is the fovea, which consists of an island of densely packed cone receptors, like a high-definition camera, allowing us to interpret fine detail. As the retina spreads out to the sides, our peripheral vision is governed by less tightly crammed receptors – fewer of the cones that detect colour and, as we move further out still, largely rods – sacrificing colour for great sensitivity, allowing us to see in incredibly low lighting conditions. But, surprisingly, as well as being non-uniform, the retina is not even contiguous; there is a large hole in the light-detecting layer of receptors, not so far away from the centre of our field of vision. The cause of this hole is entirely a problem of engineering. All those rods and cones, and those impulses signalling light, dark, colour, contrast and intensity, need to be conveyed outside of the eyeball to the brain itself. There needs to be an exit for all these signals, a portal out of the globe of the eye. And so the retinal outputs are bundled together to form the optic nerve, leaving the eye through the optic disc. As I peer into someone's eye through the tiny aperture of an ophthalmoscope, I look for the optic disc, cupped like the swirl of water in a bath plughole, the flow of optic information down the drain.

In the absence of receptor cells in this area of our eye-balls, we are rendered visually impaired, at least in the area of the optic disc, with small islands of blindness at roughly 15 degrees to either side of our centre of vision. Yet we are completely unaware of our imperfect vision. With both eyes open, it is perhaps more easily understood why that should be the case. In each eye, this blind spot is slightly lateral to the centre of vision, so the two areas do not overlap. The small area that is undetectable to the left eye is seen by the right eye, and vice versa. But now close one eye. Your vision still looks intact. The blind spot is still invisible, even without vision from the other eye to compensate. In part, this is because our eyes are constantly moving, darting around, and so the area of the field of vision within the blind spot is continually shifting. But even if you fixate your vision on one spot, the blind spot will not announce itself. Your nervous system, your eye and brain, will fill in the gaps, extrapolating what it can see around the blind spot. A line that runs across the full width of your vision, thereby in part falling across your blind spot, will be reconstructed by your brain as an unbroken line. In fact, the only time you will be aware of your blind spot is if the entire object you're looking at falls within it – a finger, for example, or, in clinical practice, the red head of a hatpin. Try it for yourself. Close your left eye and take a transparent ballpoint pen with a coloured cap in your right hand. Fixate your vision on a point in the distance, then gradually bring the pen towards the centre of your field of vision from the side, along the horizontal. As you approach the centre of your field of vision, a few inches to the right, the pen cap will momentarily disappear as the whole of it is projected on the optic disc of the right eye, undetectable to the retina or your brain. Move only a few millimetres, and the cap will miraculously reappear as its image falls back on the retina.

For most of us, this design feature of the eye is imperceptible and immaterial, nothing more than a quirk of human physiology, to be noted in biology lessons in school. But for

some people, blind spots are more meaningful, more immediate, more intrusive. If the optic disc is swollen, by increased
pressure inside the head or compression of the optic nerve, this
physiological blind spot may enlarge, so that objects are more
easily swallowed up within it, making the black hole of vision
more obvious. Even more distressing is the presence of a blind
spot in the centre of the field of vision, the fovea, where images
are focused for the most detailed vision, to read, to recognise
faces, to see in detail. Macular degeneration, a breakdown of
the retina in precisely this area, leaves people with disruption
of vision in precisely the area where you most want it; a life of
constantly trying to look around a haziness or distortion, the
tantalising view of something in your periphery, the temptation
to look directly at it only to find that when you do so you cannot
see it clearly at all. With retinal detachment, meanwhile,
swathes of the retina come away from the eyeball, leaving large
areas of vision desolate.

One overriding feature of these blind spots, resulting from
problems within the eye itself, is that they tend to affect the
vision in either eye differentially. A retinal detachment or macular damage in one eye may make little difference to overall
vision, the other eye being unaffected and still 'seeing' what the
abnormal eye has lost. But it is not just damage to the eye itself
that results in loss of vision, as we have already seen. Injury
to any part of the visual system can do this. Understanding
the organisation of these visual pathways can help identify
the location of any impairment, depending on the nature of
the blind spot someone describes. As the impulses from each
retina travel out of the eye, an optic nerve carries those retinal
signals back towards the brain. The left and right optic nerves
carry entirely separate visual information, from the left and
right eye respectively. But as the optic nerves track back from

the eyeballs, they merge, deep behind the bridge of the nose. In this area, called the optic chiasm, the first bit of integration between left and right eye occurs. Fibres carrying information about the left field of vision from the left eye, cross over to the right side, while those fibres carrying information from the right field of vision of the left eye do not; at the same time, fibres from the right eye, conducting the right field of vision, also cross over to the left. Information from the two eyes is fused and communicates the whole field of vision, with information about the left visual world carried on the right part of the brain, and vice versa. Essentially, beyond this point, the eyes are no longer separate.

So, by mapping out blind spots, ophthalmologists and neurologists can usually identify the source of visual loss. The flow of optic information through these pathways is like a wiring diagram of a house, a guide for the electrician trying to identify a short circuit. If the blind spots are in different locations in either eye, this suggests the problem is in the eye or the optic nerve – anywhere in front of the optic chiasm. If the blind spots are the same in both eyes, it implies that damage has been done behind the optic chiasm, once the channels of information no longer discriminate between the two eyes. And certain patterns of blind spots may even point towards damage at the optic chiasm itself. The two optic nerves come together, fusing in the optic chiasm, adjacent to the pituitary gland. An expanding pituitary, usually due to a benign tumour, may compress the optic fibres that cross over in the chiasm – the fibres that encode the lateral field of vision. Tumours of the pituitary will often result in loss of both eyes' lateral field only, essentially causing the vision of a blinkered horse – normal immediately ahead, but with nothing to either side.

*

THE MAN WHO TASTED WORDS

For Dawn, thankfully, the wait for further answers was not too long. She was soon given appointments with an ophthalmologist and neurosurgeon in quick succession. Her MRI scan confirmed the explanation – the presence of the tumours – but there was a wave of relief when she heard the news that the tumours were benign rather than malignant, meningiomas derived from the outer lining of the brain rather than intrinsic to the brain itself. But the good news was tinged with bad. A solitary tumour was responsible for her visual loss, wrapping itself around the optic chiasm, explaining why both eyes were affected; the tumour was gradually squeezing the life out of the optic nerves where they come together, destroying the connections between eyes and brain. The size and location of the tumour was highly challenging, but without surgery it was clear to Dawn and her surgeon that she would continue to go increasingly blind. At the age of twenty-nine, with two young children, the prospect of going blind was horrifying, and despite the difficulties of surgery and the lack of guarantee of saving her vision, Dawn decided to proceed. 'Harrison, my son, was only six at the time,' she recalls, 'so the best way we could actually explain it to him – which probably sounds quite silly – was as the TV and the video player. It was the cable in between the two that wasn't working. And that seemed to be the best way to make him understand, because we weren't sure after the surgery how much sight I would actually have.'

The family were coming to terms with the diagnosis, making decisions about risky surgery, managing how to break the news to the children. Meanwhile, Martin, serving in the British army, was about to be deployed to Afghanistan. It must have been a horrendous time, but even now, as Dawn recounts this moment in their lives, she has a levity in her voice, a no-nonsense narrative. She remembers the conversation with her neurosurgeon well. 'I did actually say that I didn't want too much information, because the choice to have the surgery in some ways had been taken out of my hands – because if I didn't have the surgery, the sight was going to continue to deteriorate. But if I had the operation, there was the possibility that they could save some of my sight. It wasn't really a question, was it?'

By the time the surgery took place, Dawn's vision was already poor. I ask her if she remembers what she could see at that time. 'Things were not brilliant,' she says. She could just about make out Martin's face, and could see the television screen if she sat very close to it. 'But I remember sitting on my hospital bed the night before the surgery, and things didn't seem too bad.' She remembers waking up from the general anaesthetic. 'I was sitting on the ward. I didn't want to ask anybody to turn the lights on and open the curtains. But apparently the lights were on and the curtains were open.' It took her some minutes to realise that she was now completely blind.

To describe the aftermath of the surgery as difficult is an understatement. She had been warned that it would take months, even a year, to fully appreciate whether any sight had been saved. Martin recalls feeling sick at the realisation that Dawn, bloodied bandage wrapping her head, could see nothing at all – no light, just dark. It was the uncertainty that they found most testing, not knowing if Dawn would ever recover, or if she would remain in perpetual darkness, and all the while trying to explain this to their two young children.

Over the intervening years, I have seen Dawn's vision fluctuate. At times she has been able to make out aspects of my face – 'I can just about see the outline of your head, and a bit of movement,' she tells me when I ask her if she can see the colour of my hair, 'but nothing more than that.' Her vision did improve somewhat in the months that followed the surgery. 'I could see a haziness in my left eye, like looking through very thick fog,' she says. But describing her vision now, she says there are 'no details, and only some colours. I can go into a shop with my daughter, Charlotte, and show her a jumper that I think is very nice; I think it is cream but it turns out to be a luminous pink.' Charlotte grimaces and laughs at this.

Over the past year or two, unfortunately the fogginess in the centre of her vision has also gone black. The vision in the right eye has never really recovered after her surgery, and has been lost for ever. Dawn is now left with a tiny island of sight. 'At the moment, I have a very small patch of better vision at the top, and if I tilt my head, get my chin on my chest, I can sometimes make out details.' The strangulating tumour has damaged the nerve fibres beyond repair, perhaps compounded by radiotherapy to the remaining tumour a year after surgery, an effort to shrink the tissue left behind after the operation. Further surgery is too difficult, simply too risky to contemplate.

In my mind's eye, Dawn still sees. When I see her in my clinic, she moves around the room with ease, looks me in the eye when we are talking, looking to all intents and purposes like someone with normal vision. But her resilience and resourcefulness mask the degree to which her vision is impaired, and in the past couple of years I have heard the tapping of her white stick as she turns into the clinic room from the corridor outside, the most obvious reminder of her blindness.

In our clinic appointments, she is not someone to dwell on the impact on her life, but outside the constraints of the clinic, in her own home, she opens up a little. 'I can now see light and dark, but no detail, no faces.' When I ask her how her other senses help mitigate the loss of sight, she laughs. 'I've become very good at burning myself when I cook!' She pauses, and continues on a more serious note. 'I rely quite heavily on my hearing, but sometimes I have quite loud tinnitus in my left ear,' – another tumour sits directly on one of the nerves supplying the ear – 'which disrupts the hearing aspect as well. There's smell, obviously. I do have to feel for things if I've dropped something on the floor – on your hands and knees, searching. Breaking things is horrific. I have broken a glass once and had to just shut the kitchen off until somebody was around to come and do it for me. Losing my independence – I think that was probably the biggest thing.'

The loss of independence while trying to bring up two young children makes Dawn's experience all the more poignant. The Army have been very accommodating. Martin's forthcoming deployment to Afghanistan was quickly shelved, and a shift in role to physical training instruction has made life a little more flexible for the family. The tight-knit community of Army families has been a life-saver for them, providing support and help, and despite the transformation of their lives over the past decade or so, the loss of vision, and Dawn's other problems that we will come to explore, as well as the uncertainty about what lies ahead, she and Martin are, amazingly, still smiling, still laughing, stoical in the face of adversity that would leave most of us crushed.

Loss. To no longer have something, or to have less of it. The loss of life, loss of a relationship, loss of a precious object, loss of money – that something you once possessed is no longer there. The pain or discomfort of knowing what it was like to hold something or someone in your hands, and to feel that absence, the ache of longing. For some people, however, the absence of something you have never possessed can also be painful. If you have never had something – an arm, for example, or a leg, the ability to walk – you can see that those around you have these things or abilities, and you can still comprehend that you are missing something. The emotional response may be different, in that you have never experienced or known anything to the contrary. However, this is loss of a different nature; your life is unchanged, constant in your experience of it, but your perception of the impact of this loss on your life can nevertheless be profound.

But when it comes to the senses, there is nothing to see, no palpable comparison. None of us knows the inner experiences of another – whether someone sees the same red as you, feels the same degree of pain as you, hears music in the same way you do. In the face of loss of a sense, the only comparator is your own experience of that sense before the loss and after, an internal control. But if you are born with the impairment of a sense, you may live with it all your life without knowing it unless it manifests in an obvious way, through a test, a mishap, a medical examination, an incidental finding on a visit to the doctor for something entirely unrelated. These occurrences are sometimes termed 'incidentalomas' – an interaction with a medical professional that identifies a 'problem' you were never aware of, perhaps in a scan showing an abnormality of your brain that has probably been there your entire life, or some subtle asymmetry of strength or movement that you have never noted.

In many ways, Oliver is a prime example of this. Now in his mid-twenties, he is forging a career as a film-maker. When I meet him, he has been busy directing music videos with a friend. His working life, the media of film and photography, is vision. He inhabits a visual world. 'I have a little point-and-shoot camera with me a lot of the time,' he tells me. 'So if I see anything that looks really cool, I can just pull it out and take a snapshot.' He is slight and softly spoken, but despite his quiet voice, I can perceive his passion for what he does, his immersion in this world of colours, shapes and movement. And with Oliver sitting opposite me, casually dressed in dark colours, I see no evidence of any medical issue, no overt sign of disease or dysfunction.

I ask him how he came to this point. He looks a little sheepish for a brief moment, then says, 'It's kind of a weird thing. I work with my friend. Basically, we were working one day, and he was quite bored because he gets distracted very easily. He started poking me in the face,' Oliver laughs, before continuing, 'And I couldn't see his hand! And he was like, "Why can't you see this?" He was just shocked that I couldn't see anything. But I didn't really think anything of it. This was maybe a year or so ago.'

A few months later, Oliver began to experience some migraines, accompanied by a visual aura. He describes the classic features – the flashing lights starting on one side of his vision, gradually spreading across his whole field of vision over a period of twenty minutes, before abating and resulting in a headache. Oliver tells me, 'I went to the doctor because the migraines were getting worse. One lasted about two hours. And while we were chatting, she just very casually did a field of vision test with her hands. And I couldn't see her left hand. She was very shocked at this, which is never a good sign when your doctor is shocked and confused.'

This chance finding triggered a diagnostic and emotional roller-coaster for Oliver. The simple act of the doctor sitting opposite him, comparing her own visual field to his. Sitting immediately opposite Oliver, she could see her hand moving on her left, but he was unable to see the same hand. Essentially, in many situations, this is what we doctors do when we examine patients; we assume that we are the 'normal' ones, assessing the patient's faculties against our own strength, our own dexterity, our own hearing, our own vision. It has always struck me as odd that we spend our lives looking for abnormalities in others by using our own bodies and nervous systems as templates, as the gold standard for neurological functioning. Given that we will frequently see patients who are unaware of the deficits in their neurological systems, much like Oliver, there is no reason to believe that we ourselves are not harbouring some anomaly or other. In no other walk of life would I consider my body a gold standard.

'So, she sent me off to an emergency optometrist,' Oliver recalls of the aftermath of his GP visit. 'He also did a field of vision test, and was also very shocked. He said I had the vision of someone who's had a stroke.' He then adds, with more than a touch of understatement, 'Which is never good to hear.'

Oliver was swiftly dispatched to the emergency department at St Thomas' Hospital, and he remained an in-patient there for two days while a battery of tests were done. This culminated in a review at the Medical Eye Unit ward round, a Wednesday morning institution that was a firm landmark of my week both as a senior house officer and as a registrar. Oliver recalls the experience: 'I went up to see – it's quite tense – a group of ten doctors in a room. Then we waited outside for their verdicts for forty minutes.' The Medical Eye Unit ward round consists of medical ophthalmologists and neurologists, consultants, registrars and senior house officers. The patient has had all their assessments and tests in the preceding day or two, and is then brought into the large room. The junior doctors present the case and the test findings. The consultants interrogate the patient, interrogate the juniors, look over the test results and review some salient examination findings, before reaching their conclusion. I recall the first few of these I attended as a very junior doctor, with a limited knowledge of neurology and a non-existent knowledge of ophthalmology. It was terrifying, but fortunately the expectation of us junior neurologists was very low indeed. The major focus of wrath, for the ophthal-mologists in particular, were the ophthalmology juniors (our neurology boss, meanwhile, was a calm and gentle soul, in whom I never caught a glimpse of anger in many years). I soon learned that the best policy for me was to stand on the side-lines, watching the junior eye doctors coming into the room like lambs to the slaughter, waiting for the massacre to start. The process was always tinged with the fear that, at any point, we junior neurologists might suddenly also become the prey.

And the outcome of all these tests for Oliver? The cause for Oliver's inability to see anything at all on his right side, while remaining completely unaware of the fact? It was nothing at all to do with his migraines. 'They told me that the cause was indeed a stroke,' Oliver says, 'but it happened when I was being developed in the womb, or straight after birth.' A small blockage of a blood vessel supplying the primary visual cortex at the tip of his left occipital lobe has resulted in the absence of the part of the brain responsible for visual awareness, of conscious sight, in the right side of his world. Oliver's vision has likely always been this way, since the day he was born or just shortly after. Without having a reason to compare himself with anyone, he has always assumed that his vision is just like everybody else's, normal in every way. He has never known any different. Of the diagnosis, he says, 'I guess there were quite a lot of emotions going on that day, because I had no idea what it was going to be. You always rush to the darkest conclusion. I didn't know whether my sight was going to disappear slowly or altogether, so, if anything, it was almost a relief to know that it wasn't going to get worse.'

It is almost implausible that someone could go through a

quarter of a century missing half their vision without realising. If it had not been for his incidental migraines, he might have gone another twenty-five years none the wiser. I ask him if there have ever been any clues, beyond the incident when his friend was poking him in the face. Had he played sports at school? 'It's probably lucky I wasn't very sporty. I'm quite clumsy.' He pauses while thinking, and continues. 'I used to work around Oxford Circus as a runner,' he recalls of his days as a junior production assistant in the film or television industry, 'and there's lots of people there. I thought I was getting good at walking around people quite quickly. But people always used to walk into me from the right-hand side. So that was a first indicator, I guess.' But Oliver also drives, and to my surprise he has never had a car accident. He does comment that, since learning of his visual issues, he has noted something about his long-standing clumsiness. 'When I think back to all the stuff I've broken or smashed in my house, it has always been on the right-hand side.' A glass perched on the table, a plate on the countertop – always on his right, in his blind spot.

I still find it difficult to comprehend how little Oliver's lack of vision in the whole right side of his world has had so little impact on his life, how there has been such an absence of overt consequences, bar a few broken household objects or the odd bump into an oncoming pedestrian. But there may be a physiological reason for this. During the course of Oliver's assessment in the Medical Eye Unit, something curious was demonstrated. While he cannot see any objects on the right-hand side – no letters, no colours, no lights – he can perceive movement. He is blind, but can see.

Britain's declaration of war on Germany in 1914 was unexpected and unprepared for. One of the many shortcomings in the national effort was that of medical provision for the conveyor belt of maimed and mutilated soldiers returning from the front. Major actions in France were accompanied by government orders to clear beds in London hospitals in antic- ipation of the bloodshed to come. Although there were field hospitals nearer to the killing fields of northern France which dealt with the immediate aftermath of injuries, officers with neurological injuries who survived long enough would be trans- ferred to the Empire Hospital in London, where one of the few neurologists of the time, a man called George Riddoch, would care for them. Riddoch was born and educated in Scotland, qualifying in medicine from Aberdeen University at the age of twenty-five, a year before the outbreak of the First World War. In 1914 he was appointed a captain in the British Army, to join Gordon Holmes, a giant of British neurology, as the second neurologist in the armed forces. Riddoch's neurologi- cal expertise was very much in demand, and in later years he became instrumental in the care of people with spinal injuries throughout the UK.

But in 1917, Riddoch observed something peculiar in his patients. He described five soldiers with blast injuries to the brain, all rendered blind due to damage to the occipital lobes. While they could not see stationary objects, they did report being able to see moving targets. Yet, although they could perceive movement, the nature of the moving objects was indistinct – 'vague and shadowy', as Riddoch described it. There appeared to be a disconnect between the conscious perception of the form of an object and its movement – a dis- sociation between the static and kinetic. This phenomenon, termed 'Riddoch phenomenon', has subsequently been repeat- edly described.

There remains some uncertainty about the precise

neurobiological basis of Riddoch phenomenon. It is generally held that we cannot have conscious awareness of vision without an area of the occipital lobe at its very tip, the primary visual cortex, or 'V1', where visual inputs to the cortex first enter. Damage to V1, as in Oliver's case, results in a deficiency termed 'cortical blindness' to discriminate it from forms of blindness caused by damage to the eye or to the optic nerve.

But Riddoch phenomenon causes us to question this classical view, that all visual information enters the brain via V1. If that were indeed the case, damage to this area would likely result in no visual inputs at all entering other visual areas of the brain, but this is clearly not what is happening in Oliver's case. Riddoch phenomenon therefore implies that there are alternative pathways, bypassing V1, that allow signals to progress to higher areas of the brain, at least for some aspects of vision. And indeed there is evidence for the existence of these alternate pathways for the flow of visual information. In essence, some, albeit limited, information about movement gets through, circumventing V1 and reaching some other visual areas, in particular an area called the middle temporal motion complex, where movement in the visual field is detected. Thus it is possible to see movement even in the context of being blind.

Why these alternate pathways should exist is a matter of speculation, but clearly, from an evolutionary perspective, it is important to know if something big is moving in the periphery of your vision, even if your vision is damaged. If you see the tiger running towards you, you may survive to pass on your genes. If you don't, you end up as lunch. And perhaps less dramatically, Oliver's Riddoch phenomenon has enabled him to avoid bumping into too many people or smash into oncoming cars while driving. He has an awareness that objects are moving to the right of him, even if he cannot really see them. And when

I examine him, doing the same field-of-vision test that his GP performed on that fateful day, it is clear that, while he cannot tell if my hand is present on the right side of his face if I just hold it there, he is aware of my hand if I move it quickly.

The concept of being able to see something while being totally blind, as in Nina's case, or in Oliver's case blind in one half of his world, is odd enough. It highlights the complexity of the visual pathways in our brains, how specialised the different areas of the brain are in processing different aspects of what we see. But vision gets even odder.

As I explained in Chapter 2, there are two major pathways for visual processing – the 'what' and the 'where', and within these pathways exist specialist areas responsible for colour, texture, faces, letters, movement, position, and so on. Normal vision (whatever 'normal' is) requires all these regions of the brain to function in synchrony, to combine to give us an accurate, or at least relatively accurate, representation of the world from which we can draw meaning and facilitate our survival.

As a young medical student, I opted to study a module entitled 'Physiological psychology'. It was at that time that I was asked to write a paper on why we dream, which triggered a lifelong interest in sleep. But the very first week in that module was dedicated to 'blindsight', this oxymoron of a term that makes no sense, akin to the sort of sighted blindness that Oliver has. The phenomenon of blindsight represents something broader than Riddoch phenomenon – that some individuals, despite having an area of the vision in which they are totally blind, or at least unaware, may still be able to see, on an unconscious level, not just movement, but also colour, form or shape. This appears to be nonsensical, since if someone can see something, then they are by definition not blind, and there have been arguments raging around blindsight since its formal characterisation in the 1970s. Essentially, the 'blind' in blind-sight refers to the fact that people report not seeing anything in an area of their visual field related to damage in their primary visual cortex. They perceive themselves as having no vision, either partially or wholly. But if forced to guess the colour, the form or direction of movement of a visual object in their blind spot, they get it right much more than would be the case by chance alone, sometimes up to 100 per cent of the time.

The true existence of blindsight has been repeatedly questioned and alternative explanations have been proposed. What does conscious vision actually mean? Is it simply that in those individuals or animals, the primary visual cortex responsible for visual awareness is not totally destroyed, that islands of functioning V_1 remain? Perhaps blindsight could be explained by light scattering inside the eye, falling onto areas of normal vision? But experimental work in patients and monkeys has largely addressed these objections, and the consensus view (although there are few areas of science on which there is unanimity) is that blindsight is a real phenomenon. In the absence of conscious vision, information about the visual world still gets through to the brain about movement, but also colour, shape, orientation and even facial expressions and the emotions associated with them. And there are multiple pathways for this flow of data, some of which work around the primary visual cortex, the seat of conscious vision. When it comes to the brain, the concept of blindness is not so straightforward.

So what does this imply for all of us? Perhaps that we are all constantly seeing without being aware, that our vision provides us with more information than we are conscious of, that our visual world is more informative, richer and nuanced than we can ever appreciate. By extension, if more information is entering our brains than we are aware of, this means we are conscious of only a part of our visual world and that what we experience is limited; it is not the whole of our environment, not a complete recognition of reality.

I am intrigued as to what this experience has done to Oliver's view of sight and the reality of the visual world, especially as someone who lives and breathes vision, who works in that medium. He describes it in a metaphor that befits him as a film-maker, referring to aspect ratios – the screen width in relation to its height. 'It brings to the fore how each person's perception of the world is definitely different. I feel like other people are looking at the world in widescreen; I'll be in a 4:3 ratio – a bit more of a traditional 1940s-film ratio, which is different rather than inferior, I'd say.' I am intrigued by his comparison of his vision to the different aspect ratios in films. It makes me think of the button on the TV remote control that changes the aspect ratio of the TV screen. When you first change it, it looks strange, distorted or simply uncomfortable, but within a few moments it is no longer perceptible. When you watch a widescreen film in a 4:3 ratio, you lose some of the periphery of the picture, your viewing experience changes subtly, but it does not intrinsically alter your ability to appreciate the film.

In fact, the more I consider Oliver's metaphor, the more I am taken with it. If you consider what we see on the screen at the cinema, it is probably radically different from the reality of the film set. It is obvious that the physical nature of the world around us is not precisely what we see. It is dependent on our body, its structure and function. An obvious example of this is colour blindness. A quirk of fate, or perhaps more precisely the inheritance of a defective gene, causes abnormal functioning in one of the three types of cone receptor in the retina, resulting in a different perception of colour. A red apple may look green to someone with red–green colour blindness, the commonest type, or may even look grey in rare cases where perception of colour is totally absent. Similarly, different animals may see the world very differently, their eyes equipped to see wavelengths of light outside our own perception. And so objects themselves have no fixed colour, but their hue or tint is dependent on their observer. Colour is in the eye of the beholder.

When we look at Oliver's film work, what we see is a function of the recording: the lighting intensity, the framing by the cameraman, the angles of filming, the settings of the camera itself – the depth of field, colour intensity, contrast. All are parallels for the machinery of the eye: the pupil, the retina, the direction the eyeball is pointing. But what we see in the cinema is not just a function of what is recorded. It is also a product of how it is projected: the colour of the lightbulb in the projector, the errant hair on the lens projected several feet high onto the screen, the aspect ratio of the screen itself. The conveying of data – visual information – from the film running through the projector at several feet per second to the white screen on the wall is like the passage of information from the eye to the brain. Perhaps even the subtle clues or subliminal messaging that celebrated cinematographers are known for are like the unconscious aspects of vision that underlie phenomena like blindsight or Riddoch phenomenon. And then there are the credits at the end, countless names scrolling down for several minutes – without any one of these people, the film would not have been the same. The film is more than just the filming. Our vision is more than just the seeing.

6

COFFEE AND CARDAMOM

'In the odor of young men there is something elemental, as of fire, storm, and salt sea. It pulsates with buoyancy and desire. It suggests all the things strong and beautiful and joyous and gives me a sense of physical happiness.'

Helen Keller, *The World I Live In*

'There is no love sincerer than the love of food.'

George Bernard Shaw, *Man and Superman*

'Every cuisine tells a story.' So Claudia Roden begins her book *The Book of Jewish Food*. 'Jewish food tells the story of an unrooted, migrating people and their vanished worlds. It lives in people's minds and has been kept alive because of what it evokes and represents. My own world disappeared forty years ago, but it has remained powerful in my imagination. When you are cut off from your past, that past takes a stronger hold on your emotions.' Roden goes on to write about her childhood in Cairo, a world of palm trees, scented jasmine and playing by the banks of the Nile, terminated abruptly in 1956, with the Suez Crisis. Ended in reality, but not in memory; rekindled by scraps of recipes kept by friends and relatives, exotic-sounding dishes with Arab, Jewish and French roots.

Every family has its food history – recipes passed down through generations, kept alive by the memories and emotions associated with eating them; dishes that conjure up the warmth and happiness of childhood, being enveloped by a mother's love. For many, particularly for immigrant families, these family recipes are also a language without words, a non-verbal history of origins, of homeland, of interactions with other cultures, as much a defining feature as the family name.

My own family's food history is even more complex than Roden's. My mother's family originated in Baghdad, the place of her birth. She grew up in a house on the banks of a big river, the Euphrates or Tigris (she was too young when she left to know which it was); a Jewish house, speaking Arabic. I recall my grandfather, some fifty years after leaving Iraq, sitting under a tree in the sunshine, drinking coffee from small thick glasses, which my grandmother had brewed in a small brass urn on the stove, while playing backgammon with his friends. The constant clacking of the wooden counters and the dice on the board. The bitter, roasted smell of coffee beans mingling with floral notes of cardamom. Guttural Iraqi Arabic back and forth, probably exactly what they would have done back in Baghdad. There would be occasional guffaws of laughter as someone no doubt uttered an unspeakably rude joke. My ears became rapidly attuned to curses in Arabic as I watched and listened as a young boy. Back in my grandparents' home, I would enter to find the table laid out with food that makes me salivate even at the thought of it: steaming saffron-stained rice with lentils, kibbeh (small dumplings of cracked wheat stuffed with lamb, cinnamon and pine nuts) or meatballs in an apricot sauce, with an accompaniment of *amba*, a spicy sweet-and-sour relish of mango, limes and a concoction of spices, a potent example of how food is history. It was probably brought to Iraq by Baghdadi

Jews who had settled in India, trading back and forth between the Middle East and the Asian subcontinent. My grandmother and her multitude of sisters would descend upon the feast, talking and cackling, while an Iraqi or Egyptian soap opera played constantly on the television in the background, the food taking them all back to a happy childhood, a privileged upbringing in what at the time was a modern, progressive city, a melting pot of cultures, languages and religions.

My father's side of the family was a complete contrast. His mother, my grandmother, had left Berlin in 1931 at the first sign of trouble, getting on a ship at the tender age of sixteen and sailing to Palestine, without her parents. She was tough, hard as nails, swapping a middle-class home in a city that at that time was considered to be the height of civilisation to live in abject privation. Her relationship with food was complicated. For her, it was simply a fuel, a necessary evil to allow her to get on with the work of the day, not a pleasure to be savoured. For those around her, it was a demonstration of love, a mode of communication. But in her eyes, it was quantity not quality that counted. The more food there was, the louder it spoke, with huge volumes of *Mitteleuropa* fare, stewed or baked to oblivion. Throughout my life, my paternal grandparents had always lived in Switzerland, and on our arrival there, without exaggeration, there would be vast piles of chocolate of all sorts and *Lebkuchen* – honey and spice cakes – stacked precariously on the coffee table. My grandmother was so insistent in feeding us that we would still be bursting at the seams from the last meal when the next was served. She took the same approach with my grandfather, constantly bludgeoning him to eat more. My grandfather was an extremely gentle man, quiet, intellectual, a polyglot with an interest in ancient history, philosophy, art and the classics, and the only time I ever heard him raise his voice was in frustration at having food forced upon him. Physically, my two grandparents were an unlikely pairing. She was heavily built, strong as an ox, loud of voice; she moved with a purpose, and there was a physicality about her. But despite my grandmother's aggressive food tactics, my grandfather remained as thin as a rake, more likely to talk in whispers, moving slowly around the house. He would pick at food as she tried to shovel it on his plate. It was only when I got a little older that I realised that his lack of enjoyment of food was something deeper, a melancholy that pervaded his life. He had been arrested on Kristallnacht in the German city of his birth, Breslau – now called Wrocław and situated in the very west of Poland – hearing the explosions as the Nazis tossed grenades into the main synagogue opposite the police station where he was incarcerated. He managed to escape the Reich on the eve of the Second World War in June 1939, he and his brother the only survivors of the family; the others all perished in the camps. The last he heard from his parents were brief Red Cross notes before they were transported to Theresienstadt. A letter from the Panamanian embassy in Amsterdam, complete with a stamp of the German eagle clutching a swastika, is all that physically remains of that time, but the trauma lived with him till his dying day. The letter confirmed safe passage to Panama, exiting Germany at that time conditional on having a place to go. For his whole life he had been an atheist German, speaking *Hochdeutsch* (High German), dressed in a suit, listening to German classical music, reading the great German writers and philosophers – yet rejected by his country of birth, persecuted, almost exterminated. Ironically, he and his wife were always known as 'the Germans' in whichever small corner of Switzerland they lived, and when he died at the age of ninety-two, he remained an island of Germany in a Swiss sea, living only a few hundred metres from the German border, which was literally visible from his home. From his front door, the burbles, whirlpools and ripples of the Rhine's clear waters were almost audible, a liquid boundary between him and his homeland; as close to Germany as he could possibly get, without actually returning to it. I wonder now what the river was murmuring to him, what he heard in that whispered conversation. A few years before he passed away, my grandfather lost his sense of smell. I recall his meagre appetite dissolving away and how this already thin man became so fragile that he looked as if he could snap at any time.

As you can imagine, the food of my childhood was highly varied, and each dish reminds me of different aspects of my family history – its origins, its traumas, its migrations, as Claudia Roden describes. But on an even more personal note, if I walk past an Arab restaurant on the streets of London and I catch the smell of Arabic coffee tinged with cardamom, I am drawn straight back to my grandfather playing backgammon under a tree, steaming glass in one hand, two dice in the other. With veal and mashed potato, meanwhile, or a large bowl of Bircher muesli, I am back in my grandparents' living room in Neuhausen, the audible rumble of the Rheinfall waterfalls in the distance. The smells and tastes are a direct road to the past, my own and those of previous generations. As for my own children, now being fed Punjabi food by my wife's mother, and schnitzel, bagels, baklava and meatballs with apricot sauce by my own parents, their food history will be bewildering and fascinating in its complexity.

During a six-month stint in oncology as a junior doctor, I would regularly see patients whose oral mucosa – the mucous membranes lining the mouth – were inflamed and painful, poisoned by the drugs we were dripping into their veins. Chemotherapy agents target the mechanisms allowing cells to divide and multiply, therefore damaging rapidly growing tumours, but other rapidly replicating tissues would become collateral damage in the pharmacological battle – tissues like the skin, which resulted in irritated and painful palms and soles; the bone marrow, causing anaemia and low white blood cells; and the gut, sometimes associated with torrential diarrhoea. And with mouths sore and tongues swollen, their sense of taste would be obliterated, gradually returning as they recovered. But outside these specific circumstances, in a clinical setting, disturbances of taste are very rare indeed. Other potential causes include medications that may cause changes to the functioning of the taste buds or alter saliva production. For example, there is one sleeping pill that seems to have a predilection for giving people a horrible metallic taste that lingers in their mouth.

I can think of only a handful of people I have seen in my clinic with a primary complaint of loss or alteration of taste, specifically. One of those is Irene. The first time we 'meet' is at the start of the second wave of Covid-19. A sudden spike in cases leads us to carry out most of our consultations by video-call or telephone – except not everyone gets the memo. I get a phone call from the hospital while I am at home, getting Zoom fatigue from yet another video clinic. Irene has an appointment, but no one has told her it is now a remote consultation. She is seated in the waiting room, without a doctor to be found. So our first meeting is over the telephone, she understandably unhappy that she has come into hospital for no good reason, me at home embarrassed by the administrative cock-up. The following week we meet face-to-face, although even then I spend the whole appointment in full PPE, with visor, mask, gloves and apron, and she wears a mask covering her face for all but a few minutes while I examine her. She has lived in London for a while now, but is originally from Toledo in Spain. Her accent is thick, and from behind her mask it is sometimes difficult to fully comprehend her. She looks typically Spanish, is young (twenty-nine years old) and, in contrast to many of the patients who I have seen that morning, looks pretty relaxed about venturing out in the midst of the pandemic.

Irene has been referred from the oral medicine clinic. For the past five months she has had no sense of taste – under the current circumstances, when she first noticed it she was sure she had contracted Covid-19 without any other symptoms. Perhaps more concerning is the fact that Irene is a sommelier. And not just anywhere – Irene has been working in Michelin-starred restaurants for several years now. Her palate is her vocation, the skill upon which all depends.

I ask Irene if she always had a refined palate, if this is what drew her into the world of wine. But it was by accident. 'I was cooking with my grandma when I was five years old. I always liked to cook at my house. My family come partly from the south of Spain, partly from the north.' The north prides itself on its gastronomy – fantastic seafood, the cuisines of Galicia and the Basque country. In the south, the pig is king, with *jamón ibérico de bellota* being the culinary gold. It comes from Black Iberian pigs raised in the oak forests of Andalusia and Extremadura, snuffling between the trees for the acorns that give the meat, cured for three years, its nutty, creamy magnificence. I am salivating as we discuss the food of her home. 'I did art for two years – the history of art, painting and that – but I was really sure that I wanted to be in a restaurant. The idea was to be a chef, to be honest.' Irene initially undertook a waitressing course, an entry point to the restaurant business, then a diploma in hospitality. Rotating through the kitchen, management and front of house, she soon realised that behind the kitchen door was not for her. 'At that point, I did say, "Okay, why do I need to be in the kitchen if I really want to be with people?"' An internship in a one-Michelin-starred restaurant in Marbella turned into a four-year stint, and she decided that what she really enjoyed was the combination of food and wine. Following a formal diploma, her knowledge of wines has led her to establishments across Spain, and for the past two years to various restaurants in London, places I can only dream of eating in.

I ask her if her palate has been learned or if it is something innate. 'I think it's something that you have, but it's true that it's important to train. It's also a skill that you develop.' She talks me through how people relate wines to flavours or aromas that they are familiar with, so you need to be conversant with a range of tastes and smells that may not necessarily be the case without training. Having spent much of her time in Malaga province, at the southern tip of the Iberian peninsula, she was fluent in the language of fruits, especially tropical ones like passionfruit or *caqui* – persimmon; she was less acquainted with aromatic spices like cloves and cinnamon.

It was between lockdowns, in the summer of 2020, that Irene noticed something strange. Restaurants had just re-opened following a lull in coronavirus cases, and she was back at work. 'The taste, it was different on one side of the mouth from the other.' I ask her to describe the nature of that change, but she struggles a little. I am not sure if it is because English is not her first language, or if it is simply very difficult to put into words. 'The sensation, it was like a bit more paste-y on the right side.' I tell her I don't know what she means by that. She pauses to think, then continues. 'The sensation of food and the sensation of the drink. It was more . . . I mean, I don't know how to express it. It was creamy. The sensation was creamy at the beginning.' She clarifies that it was the mouth-feel that initially altered, and compares it to that slightly woolly mouth-feel after eating Pringles.

Her first move was to take a Covid-19 test, which thankfully was negative. After a few days, she noticed some discomfort around her wisdom teeth. A trip to the doctor resulted in a prescription for antibiotics for a presumed dental infection. Despite the pills, her symptoms were worsening day by day. She began to experience a lack of feeling in the right side of the mouth, her gums, her cheek, and gradually across that side of her face. With a rising sense of anxiety, Irene had numerous phone calls and trips to the doctor. 'I was so stressed at that point. For two weeks I couldn't taste the acidity in wine. I felt, *I cannot have my life. I don't have my taste.* I mean, I'm waiting for a promotion now. I wanted to do more stuff related to wine. No one knew what was causing the problem. No one knew when it was going to pass. So, I was like, "Give me something to have my taste again!"'

Her persistence culminated in various dental appointments, and finally an appointment in a maxillofacial clinic with a surgeon. And this is how she has ended up sitting in a neurologist's clinic room. The surgeon organised an MRI scan of the face, to look for disruption to the nerves that convey taste. While these nerves look intact, something else is evident. The scan shows a patch of signal, white and fluffy, in the brainstem. It looks like inflammation, taking out an area containing nuclei that are the inputs for taste and for sensation from the face. A clear explanation for Irene's symptoms.

While there is an obvious indication as to where the problem originates, the cause of that inflammation is not clear; inflammation in the central nervous system has a myriad of possible causes, including various autoimmune disorders or the aftermath of a viral infection. The most likely cause, however, particularly in a young woman, is multiple sclerosis, or MS. The hallmark of this condition is the development of patches of inflammation in the central nervous system, an out-of-control immune system attacking elements of the brain and spinal cord. The precise nature of this autoimmune disease is not fully understood, but what we see on MRI scans is the development of small areas of damage that correlate with 'relapses', the sudden development of new neurological symptoms like visual loss, weakness, numbness or loss of bladder control. The auto-immune assault on the nervous system is directed towards cells that generate the myelin sheath, a thick wrapping of protein around neurones, like the rubber coating of an electrical wire. The myelin speeds transmission of electrical impulses and helps maintain the health of the underlying neurone, the wire itself. And so, the loss or damage of myelin results first and foremost in a slowing of conduction speed, but also subsequent injury to the neurones conducting these impulses. Depending on the precise location of these areas of damage, and the circuits dis-rupted, different symptoms can arise.

When I was training as a junior doctor, we would very regularly see patients whose bodies and nervous systems had been ravaged by the condition, left paralysed or blinded by the uncontrolled inflammation gnawing away at their brains and spinal cords. Even now, over twenty years later, as I drift off to sleep I remember the faces and voices of certain MS patients I have treated during my career. One such patient embedded in my past remained on the ward for the entire six-month period of my first neurology post. Her devastating MS and associ-ated immobility had left her with bed sores so deep that her sacrum – the bony structures of her lower back – could be seen in the depths of her wounds; months at home with inadequate care had resulted in the death of these tissues due to a lack of movement in bed. I was tasked with examining her sores twice a week, nurses rolling her onto her side, me peeling back the layers of dressings to her moans of pain, to reveal the gruesome sight and overwhelming stench of decaying flesh. Despite meticulous care and regular turning in her hospital bed, the sores refused to heal. The awfulness lives with me still, and when I think of her I catch a faint whiff of that terrible smell. She passed away from uncontrollable infection shortly after I moved on to my next post.

But over the years, we have begun to recognise that there are many people walking around with MS without any knowledge that they are anything other than healthy. Small, silent areas of inflammation are visible on brain scans but invisible to the individual or a doctor. And this area of neurology has changed beyond recognition in the past two decades. The old aphorism that neurology is a field with a thousand diagnoses but only one treatment – steroids – is simply not the case any longer. A myriad of treatments – injections, intravenous infusions and tablets – now exist, to suppress the immune system's attack on the nervous system. In some cases these treatments can completely halt MS in its tracks, while in others they slow its activity and progression. Thus, MS has become, to some extent, an eminently treatable, if not curable, condition – much like HIV, which was previously a terminal diagnosis but is now largely a condition limited to the outpatient setting, controlled with anti-viral medication. That is not to say that people do not experience significant disability with MS, but for an increasing number this is not the case.

Irene seems surprisingly relaxed about all this, and MS has already been raised as a possibility. She is more concerned with getting back her sense of taste. I tell her that we first need to exclude other possibilities, and I send off a barrage of blood tests. The scans show a solitary area of inflammation, but a closer look reveals a couple of tiny areas of inflammation elsewhere in the brain that may represent the telltale signs of MS. We organise a quick repeat scan, and no further inflammation has occurred in the interval, the original patch of abnormality maturing as I would expect following a single episode of inflammation. From a diagnostic perspective, all we can do is watch and wait, to see if there is any evidence of further inflammation, either through the appearance of new symptoms or a repeat scan. Sometimes time is the best diagnostic tool.

From a symptomatic perspective, however, there are signs of some improvement in Irene, not because of any specific treatment but due to masterful medical inactivity. The numbness has improved, although she still finds acidity and bitterness very difficult to taste. What I find remarkable is that Irene is still working as a sommelier, still tasting wines, still pairing wines and dishes novel to her. I ask her how she can do this if her taste is so disturbed. 'To be honest, before this happened, my ability was more related to the taste in the mouth, rather than the nose. Of course, the mouth and the nose are related. Obviously, if you don't sense in the nose, there's no sense in the mouth at all. But I was more trusting in my palate rather than my nose. Since I have had this, I have needed to train my nose.' I ask her what impact it has had on her ability to make new wine–food pairings. I am not expecting the answer she gives me – that there has been relatively little. 'The pairings are a bit subjective, no? I give the wines to my colleagues to taste and ask what they think about it as well. I taste with them, and they say, "Wow, it's great. It's amazing," and I'm like, "Okay," so that works for me as well.' Later, she tells me, 'I am a person who adapts to everything.'

It appears that Irene has adapted to her disturbance of taste, in part due to her knowledge and memory of the tastes, flavours and aromas of the various wines, but also due to an increased reliance on her nose.

There are two striking aspects to what Irene tells me. The first is how little impairment she has from of a change in one of her five senses, especially given her absolute reliance on this sense for her job. If a disturbance of taste has such little impact on her, perhaps it is not surprising that so few people come to my clinic complaining of taste disturbance. She may notice it, but it is hardly life-destroying. Perhaps for us mere palate mortals, a subtle change in taste barely registers. Her impairment of taste seems an order of magnitude less important than that of my grandfather, whose appetite and enjoyment of food vanished alongside his sense of smell.

The second aspect is the way Irene discriminates between her palate and her nose. She is clear that before her medical problems materialised, she relied much more heavily on her palate, on her mouth rather than her nose. Pushing her nose into the wineglass, inhaling a lungful of wine-laden air through her nostrils, was a minor part of the sampling process to her. But a basic understanding of the science of flavour shows this to be very far from the scientific truth . . .

At fifteen, Abi is only three years older than my eldest daughter, but I am struck by the difference. They are about the same height and both have long, dark hair, although Abi's is straight and worn loose. But from the way she speaks, Abi has clearly crossed the Rubicon and stepped firmly into the threshold into adulthood, no longer a child. Despite this, in medicine, the boundary between paediatrics and adulthood is like the Berlin Wall: immovable, fixed at eighteen years of age. In the world of medicine, she would be seen by a paediatrician, yet she appears distinctly adult to me. As I sit opposite her in her family's sitting room, it seems bizarre to consider Abi as having more in common with a six-month-old infant than an adult in their early twenties. She lives in a lovely corner of Dorset; the house is opposite a small playground and surrounded by fields. Even on a day like today, with horizontal rain, cast-iron skies and a biting wind, the beauty of this green corner of England is evident on my journey to her home along winding country lanes hugging the undulating landscape. As we chat, Abi's mother, Dawn (not the same Dawn as in Chapter 5), comes in bearing coffee, and I gratefully warm my hands on the steaming mug.

Abi has lived in the South West for as long as she can remember, but Dawn is an interloper, originally from Birmingham, and her rounded vowels are a stark contrast to the local accent. There is obviously contagion of pronunciation, and Abi's speech patterns swing wildly between a West Country burr and a Brummie twang, quite disconcerting at first but soon very pleasing to the ear.

Dawn settles next to Abi on the sofa. 'When she was very little, she was a good little toddler,' she smiles as she looks across at her daughter. 'No real issues. The only problem we ever had was at breakfast time or dinner time. She would play up nearly every meal time; she'd be really naughty.' Dawn recalls her and her husband's frustration, trying all sorts of parenting techniques to get Abi to eat her food, every meal turning into a battle of wills, adult versus toddler. Dawn continues: 'It took some years to realise that she just wasn't interested in food. As soon as she sat down at the table, she started yawning and stretching, which we knew was, "I'm not eating what you put in front of me".' I ask if Abi was interested in desserts. Dawn laughs, and says, 'Sweet stuff – she'd be very interested; looking back on it, anything really sweet, or really salty. In those days, she wanted everything really well seasoned. When you've got a toddler, the last thing you want to do is put loads of salt and pepper and things in their food. And we really struggled with that.' She recalls going out for a meal with friends, and someone ordering gammon (I think the last time I touched this was at school, and I recall chewy pink meat, the texture of a shoe sole, salty like the sea). Abi was given a small piece to try. 'And that was it. All she ever wanted for dinner was gammon. Everything was gammon!' Abi soon discovered the salt shaker, and by the sound of it the battle to get her to eat was then replaced by the challenge of preventing her from dousing everything in salt.

Abi was about three years old the first time Dawn took her to the doctor about her eating habits. Dawn was told, 'She's a toddler. That's what they do. She'll eat when she's hungry.' Dawn tried being extremely strict – 'If you don't eat your dinner today, you'll have to have it for breakfast in the morning' – the stress of needing to make her child eat leading to a constant battle. 'When I look back on it, I think what an awful parent I was.'

The lightbulb moment came when Abi was about four years old. Dawn recalls walking down a lane, taking Abi and some friends to school. A tractor hauling silage drove past the group. 'And it absolutely stank,' Dawn tells me. 'It was awful. And I looked at Abi and I realised there was no facial expression. I looked at the other children around us and they were all gasping and holding their noses. Abi just looked very blank, and it was a realisation. I asked her to breathe in, and she just looked at me as if to say, "I don't know what you want from me."'

Later, she discussed this episode with her husband. It dawned on them that there had been clues earlier on. 'When children go to the toilet and they start having smelly poos,' she chuck-les, 'and they say, "Oh, that stinks" – we had never had any of that. Not once.' And they realised that Abi had never told her parents that she was hungry, that she wanted to sit down and eat. At this point, Dawn was still questioning if she was simply looking for an excuse for Abi's apparent bad behaviour, but soon realised that this was not the case. The doctor suggested doing a taste test, so back at home Abi's parents put a blindfold on her and gave her different foods that she would have been familiar with. 'And she just didn't recognise anything. We gave her strawberry yoghurt and she just had no idea what it was. There was no link to her brain. But as soon as we took the blindfold off, she said, "Oh, it's my yoghurt."'

I turn to Abi, who has been listening quietly to all of this, stories of her childhood too early to recall herself. I ask her when she first became aware of something being not quite right, a difference between her and other children. She recalls her class walking down the school corridor, by the kitchens, and her friends excitedly discussing the fish and chips on the menu that day. 'I would be saying, "How do you know that? How do you know what is cooking behind the kitchen doors?"' Her memory is one of puzzlement when they told her they could smell it. On another occasion, someone had left a tuna sandwich in the car that they'd taken home from a children's birthday party. After two hot summer days, the sandwich fermenting unnoticed, the stench was unbearable when it was discovered. Unbearable to all except Abi, that is. 'Everyone was saying, "Oh, it is so horrible!", and I didn't understand what they were talking about.' I can only imagine how confusing it must have been for a young child to realise that there was a world out there to which all were privy except her.

For Abi, the apparent lack of taste and smell explains much

of the behaviour that her parents noted – the ambivalence to food, the lack of appetite, the failure to respond to the stench of silage or the saliva-inducing wafts from a kitchen. But it is clear that Abi's sense of taste, in its purest sense, is unaffected: the saltiness of gammon, the sweetness of chocolate, the sourness of lemon, the bitterness of coffee – all are perceptible to her. But the nuance of flavour is completely absent – 'Chips taste horrible, just plain to me. I need a lot of salt, vinegar, pepper, just to give it some sort of taste. Chocolate fudge cake or a normal chocolate bar have the same taste to me, just a different texture' – like listening to an orchestra with earplugs in, the richness of 'taste' is bled out.

And therein lies a clear truth of our gustatory world – that when we talk about taste, what we actually mean is flavour. A truth that we all experience but may not necessarily be aware of. When we have a cold, food is robbed of its flavour, rendered bland by nasal congestion. Or there is the taste test beloved by science teachers, of eating a small chunk of apple or raw onion, where the flavour becomes difficult to discriminate with the nose pinched closed. In fact, Abi has her own version of this test that she uses to try to illustrate her world to friends – the jelly bean test. She will blindfold someone and give them different-flavoured jelly beans while they are holding their nose. Their mouth will be flooded with sugar, the taste of sweetness bathing their tongues, but it is only on releasing their nose and breathing through it that the overwhelming flavour of strawberry, orange or lime will kick in.

Flavour is the ultimate illusion. What we perceive as one sense – taste – is not. It is not even two senses, but three. It is the perfect confluence of taste from the taste buds of our tongues and the olfactory receptors of our nose, but also the sensation of the texture of food – so-called 'mouth-feel'.

In the absence of smell contributing to Abi's experience of flavour, she is unsurprisingly more reliant on texture to guide her likes and dislikes. 'I really like red pepper. It doesn't really have a taste to it – it's sort of watery – but it's got a nice texture. That's why I don't like mushrooms – because they're slimy. It makes me think of eating a slug. It's disgusting. I base taste on what I can see and what it feels like, as opposed to the actual taste.' She describes the experience of eating a burger and her language is fascinating; she talks about the heat of spices, the crunchiness of the lettuce, the runniness of the mayonnaise and the softness of the bun, but no concept of flavour.

In essence, taste is a primitive basic sense, with limited detail or resolution. Our eyes allow us to experience an almost unlimited palette of shades and hues, our ears grand symphonies or the drop of a pin. In stark contrast, we can only discriminate five (possibly a few more) tastes. Historically, we only talked of four tastes – sweet, salt, sour and bitter – but the relatively recent discovery of distinct taste cells sensing glutamate – what gives savoury richness, or 'umami', and specific nerve fibres transmitting these impulses, has added a fifth primary taste modality. In the past decade, a potential sixth taste has been recognised, too, with fat receptors identified as recently as 2019. So perhaps in the next few years we may talk about six primary tastes, but regardless of the ultimate number – be it five, fifteen or fifty – this is a hugely simplistic way to perceive the world when compared with the hundreds of different types of human olfactory receptors allowing us to discriminate many thousands of smells.

The fundamental organ detecting taste is the taste bud, a cluster of sensory cells embedded in the tongue. On a microscopic level, a taste bud looks like a garlic bulb buried in soil, the very top of the bulb breaking through to the air. Multiple elongated cells within the bulb are like the individual cloves of garlic, their tips reaching the surface of the lining of the mouth. These cells are not identical, and indeed respond to different tastes – some, for example, only to sweet, others only to bitter – while each taste bud, consisting of an assortment of taste cells, will respond to a variety of different tastes. And contrary to what we have all been taught – that different parts of the tongue detect different tastes – the reality is that we can detect all tastes in all areas of the tongue. And this is not the only taste myth. Another is that we only taste with our tongue. In fact, taste buds are found in our palate and even the epiglottis, the flap of tissue that closes when we swallow, to prevent food or liquid going down our windpipe.

The nature of these cells, and what it is they taste, is under genetic control, and therein lies an explanation for some food oddities. Genetic variants have been associated with the perceived intensity of sweetness or ability to detect umami. Perhaps the best known of these genetic influences, however, is the hatred of Brussels sprouts and broccoli. Owing to a genetic variant, some individuals find a chemical called phenylthiocarbamide (PTC) and related substances extremely bitter, others only moderately bitter, while another group cannot taste these compounds at all. Individuals carrying the genetic variants resulting in being a bitter 'super-taster' may have a genetic advantage – they may find it easier to detect and avoid certain dietary toxins. But there may also be negative consequences. The avoidance of sprouts, broccoli and other cruciferous vegetables may also be hazardous, limiting the intake of the anti-cancer phytochemicals that these foods are rich in. This has led some researchers to closely examine the association between these genetic factors influencing taste and risk of cancer. Most of these studies have not found any association, though one study did find a link between taste genes and risk of colorectal cancer. It's certainly not enough evidence to convince children to eat their broccoli, but is a curious reminder of the huge influence of the make-up of our sensory apparatus on our experience of the world, a quirk of genetic fate influencing whether we can detect a chemical in our environment or are completely ignorant of it – a form of taste colour-blindness.

So, as Abi and her jelly-bean test illustrate, when we use the term 'taste' it has two entirely different meanings. In the physiological sense, it refers purely to the act of detecting chemicals of specific types dissolving in our saliva and triggering receptors on our tongues and in our mouths. But in a human sense, it refers to the experience of flavour, the subtleties of food and drink that enrich our lives, an integration of taste, smell and feel. And without the nose, there is no taste in the broader sense of the word. When we smell the world around us, we draw in air from our surroundings, sniffing a delicate rose 'orthonasally'. But as we chew food, or drink swills around our mouth, we smell in a different way. Volatile compounds from our oral cavity are drawn 'retronasally' up beyond the hard and soft palate, then pushed forward into our nose. So, while our taste sensations are very limited, our experience of flavour is created somewhere in the depths of our brains, an illusion of our minds based on the taste, retronasal smell, temperature and texture – not one sense but many. This goes some way to explaining why Irene's ability to sample wine is only modestly affected. While she has lost limited information about acidity or bitterness, her palate is largely defined by retronasal olfaction.

On an experiential basis at least, then, taste, or should I say flavour, is an amalgam, like an alloy; indivisible, qualitatively different from the sum of its parts. But from an evolution-ary perspective, taste and smell remain separate. The sense of taste can be considered a nutrient sensor or a detector of toxins, identifying foodstuffs that will keep us alive, help us grow or be the death of us. Sweet equates to sugar as a source of energy. Umami is the taste of protein, the building blocks of life. Bitter is the warning of poison. Smell and mouth-feel add shades of colour, distinguishing the floral sweetness of a cold, crisp strawberry from the rich, creamy sweetness of chocolate.

Flavour is almost as important as taste in our evolution. Our responses to sweet or bitter tastes are predictable and stable – reliable signals that a food contains energy or toxins – but, in stark contrast, our responses to flavours are malleable and can be learned. We have all had the experience of eating some-thing that has made us very ill the following day. It is highly likely that you will have avoided that food for months or even years afterwards. A single unpleasant exposure can put you off a certain food for life. But if you become ill from something sweet, to avoid anything sweet for the rest of your life is defi-nitely not conducive to survival and passing on your genes. It is flavour – essentially largely retronasal olfaction – that deter-mines our liking of a particular foodstuff, or our revulsion. It has the additional benefit of allowing us to learn to like what is available if it has nutritional value, even if to some degree it may taste, say, slightly bitter. Examples of this are abundant – substances that children inherently dislike but that we learn to love as adults, either due to some nutritional benefit or because in some other way they make us feel good: beer, whisky or coffee, for example.

On an anatomical level, too, the components of flavour are distinct. The pathways of information into the brain are separate. Smell, through the olfactory bulb and nerve, is the only sense with such a direct path into the higher – more complex – areas of the brain. In contrast, signals from the taste buds pass into the brainstem in a convoluted manner. From the front of the tongue, taste fibres run within the facial nerve, which is usually considered a motor nerve, mediating movement of the facial musculature. In fact, one of the commonest causes of change of taste in my neurology clinic is in people with Bell's palsy, the sudden paralysis of one side of the face caused by inflammation of the facial nerve. But taste is also conveyed by other nerves: the glossopharyngeal nerve, running from the back of the throat and tongue, and the vagus nerve, from the palate. These signals convene at the gustatory nucleus in the lower brainstem, the region damaged in Irene, before being relayed up to the cerebral cortex. And it is here, in the upper reaches of the brain, that flavour is created – in a network of areas centred on the insula but involving many areas of the limbic system, influencing emotion, memory and reward – where taste, smell and oral sensation coalesce.

But this concept of flavour being a product of the brain, not of the food – that flavour is an illusion – is a feature at an even earlier level in the neurological process. As early on as the mere detection of stimuli in the mouth and nose, illusions abound. When it comes to our mouth, confusion reigns. The place in your mouth that you taste something is related more to mouth sensation than to taste; it is touch that determines the location of taste. So if I put a drop of sugar solution on the right of your tongue but place a swab on the left of your tongue, you will taste the sugar on the left, not the right. Equally, if you taste something in keeping with an odour, you are much more likely to perceive a smell as arising from the mouth, considering it a taste. Ultimately, each of these inputs influences the perception of flavour – touch captures taste, which then captures smell, resulting in a unified experience.

The causes of loss of smell are many. At medical school, we are taught about the ones that are rare but important: a meningioma – a benign tumour – in the olfactory groove, disrupting the flow of information from the olfactory receptors to the brain; genetic disorders that prevent the development of any apparatus of smell – you may recall that Paul, in Chapter 1, has no sense of smell due to the mutation that also leaves him unable to feel pain; a serious head injury, causing a shearing of the small nerve fibres as they pass through the base of the skull into the nasal cavity; or, as I have written about in my previous book *The Nocturnal Brain*, as a precursor to degenerative diseases of the brain, such as Parkinson's disease. But in reality, other more prosaic causes are much more likely: disorders of the nose causing inflammation or blockage to airflow; viral illnesses directly damaging the nasal mucous membranes or the olfactory neurones themselves; as I write, loss of sense of smell is ubiquitous as a feature of Covid-19, a red flag to self-isolate and seek a test.

The realisation that the novel coronavirus could cause the loss of smell and taste sent shivers through the world of neurology. Historically, neurologists have been at the forefront of several epidemics over the past century, viruses having an affinity for infecting and damaging the nervous system – infections like polio, causing paralysis in the young. We frequently visit our long-term ventilation unit at St Thomas' Hospital, seeing patients with chronic conditions of the muscles or nerves causing weakness of the muscles used in breathing. Until only a few years ago, at the entrance to the unit there was a huge, steel, coffin-like apparatus, like a one-man submarine, fixed to a bed frame. On either side were chrome ports, like nautical port-holes, where hands could be inserted. At one end was a flexible seal, for the head to stick out of, and within this monstrous machine the rest of the body would be encased. This 'iron lung' was the mainstay of keeping patients with polio alive, fluctuations in pressure within this tube causing movement of the chest wall when muscles could not, rendered weak and useless by the virus. Photos from the 1950s show huge, warehouse-like wards filled with row upon row of iron lungs, heads of the afflicted sticking out at one end, mirrors perched above their faces so they could see around them. There were other neurological epidemics, too, like encephalitis lethargica, which killed about 1.6 million people over a decade during and after the First World War. This mysterious illness, perhaps related to the Spanish flu epidemic, disappeared as mysteriously as it came, leaving behind not only the dead but also some 5 million survivors ravaged by the condition, with symptoms similar to Parkinson's disease as well as profound sleepiness and psychiatric problems. And so the fact that this new virus, Covid-19, also seemed to have a predilection for nerve cells – the olfactory receptor neurones – raised the spectre of an impending neurological catastrophe. We had visions of explosions of encephalitis, paralysis and other horrendous complications. Fortunately, these apocalyptic prophesies have not been borne out. While we have seen some neurological complications of Covid-19, the numbers of people affected have been relatively small, and the loss of sense of smell seems more likely to be due to infection of cells within the olfactory epithelium rather than the neurones themselves. While a number of viruses may enter the brain via this route, it appears that, luckily, Covid-19 is not one of them.

This gateway that the olfactory system represents, from the

outside direct to the central nervous system, may, however, have some broader implications. In addition to the loss of sense of smell in the prelude to Parkinson's disease, hyposmia – the clinical term for reduced smell – is a feature of other degenerative disorders of the brain, such as Alzheimer's disease. This may of course be related to changes in the areas of the brain responsible for smell, but evidence points to the olfactory nerve, and the receptor neurones in the nasal cavity, also being affected. It may be that in people at risk of Alzheimer's, or indeed those who have it, those stem cells responsible for the regular regeneration and replacement of these nerve cells are less active. But the association of hyposmia with these neurodegenerative disorders, as well as other findings, has led to a fascinating hypothesis as to how these disorders may start. Specific viral DNA has been found in the brains of patients with Alzheimer's disease. The 'infectious hypothesis', as it is known, proposes that the production of abnormal proteins resulting in nerve-cell damage is facilitated by certain viruses from the herpes family, and that these viruses enter the brain through the nose. The theory suggests that these abnormal proteins, known as beta-amyloid, may have antimicrobial functions, so the protein tangles that are the hallmark of Alzheimer's constitute a response to viral infection.

In Abi's case, however, the cause of her inability to smell remains a mystery even now. A scan has revealed an olfactory bulb that is smaller than normal, but it had already been noted at an early age that she had some issues with her nose – her mother noticed that she snored a lot at night and an assessment revealed extremely large adenoids. Perhaps these were blocking the flow of odour from the mouth up to the nose, interrupting retronasal olfaction. An operation was undertaken to remove her adenoids, to no avail. At the age of around eleven, a further procedure was performed to correct a deviated septum and to address some chronic inflammation. Abi recalls visiting family in Birmingham shortly after this second operation. 'I was playing on the trampoline with my little cousin. He accidentally whacked me in the nose.' Later on, Abi went inside, and went to feed her aunt's fish, a favourite activity of hers. 'I always loved feeding the fish as a kid,' she says. 'And I smelled the fish food. I couldn't link it to anything. It was something that wasn't just air, if that makes sense. I wouldn't say that I could *smell* it, because I had no idea what it was. It was just like if you're blind, for example, and all of a sudden you just see a little white flash. And then it was kind of gone.' For a few weeks subsequently, Abi thought she might be able to detect the flavour of orange in her chocolate-orange biscuits, rather than simply the sweetness and texture of chocolate and orange jelly, but that, too, disappeared. Like a flash of light in the sky, when you are unsure whether you saw a shooting star or it was simply a trick of your imagination, Abi is uncertain if she has ever 'tasted' orange. She wonders if her cousin's elbow to the nose may have shifted something, perhaps dislodged some scar tissue.

But Abi remains without smell, and is acutely aware that a part of life is totally closed off to her. On one occasion in school, her homework was to write a story based upon the five senses. Abi protested that she could not do it as she had no concept of smell or taste. 'So the teacher said, "Talk to your friends about it, then." But, how could I talk to them?' She was being told to ask a question, when she could not even grasp the language. 'I've asked people to describe smells before, and their response is, "It is what it is." People tell me that grass can smell. That is mad to me. Grass is everywhere. How can you even recognise it as a smell? And when I ask friends to describe it to me, they say, "Well, it smells like grass, doesn't it!"'

Abi's frustration is evident as we talk about this, her inability to comprehend what is woven into the fabric of life for everyone around her. For most of us, no interpretation is needed. Our sense of smell is instinctive and unconscious. It is this immediacy that perhaps explains our limitations of language to describe smells, in contrast to those qualities of our other senses.

For Abi there are constant reminders – her friends talking about the food in the school canteen, the smell of the classrooms, the memories stirred by tastes and smells of food. 'It is so constant that it is not noticed, not appreciated, I think' – by everyone except her. And the role of smell as a signal of danger also has implications for Abi. A recent incident involved something burning in the tumble dryer. 'We had billowing smoke on the ground floor, but Abi was up in her bedroom, not aware of it at all,' says Dawn. The house is now fitted with connected fire alarms and gas detectors. An inability to smell if food is past its best means Dawn has to constantly remind Abi to check the use-by date on packaging. Social dangers lurk as well. Another of Dawn's parenting duties is to monitor for body odour. 'I would say, "Let me smell! Don't forget to put your deodorant on every day!"' Essentially, Dawn is Abi's nose.

I ask Abi and Dawn about their worries for the future in a world without smell. Dawn starts, 'I worry that I won't be here to check that she doesn't smell, that her food is not off. And she'll get boyfriends and she will want to wear their shirts and to know how they smell.' Abi adds, 'My best friend, she says, "I love it when he gives me a hoodie, because it smells like him." Why would you appreciate that? Wouldn't it just be sweaty? But apparently it's a really nice thing to have. It links them together almost. And I'd like to have that when I'm older.'

Beyond the role of smell in the signalling of danger, in flavour and, as Abi implies, in memory, there is another aspect of this sense that is undervalued – that of human interaction, of guiding social and sexual behaviours. The presence of volatile compounds – odours – in body odour, urine, faeces or blood, can precipitate fear and vigilance in prey species – a warning of a predator in the environment. The detection of chemicals marking out sick animals results in others avoiding them, the smell of illness a powerful survival skill. While this is appreciated in other mammals, such as dogs, in primates like ourselves there has been an evolutionary reduction in the importance of the olfactory system as it became increasingly expendable in the face of the development of vision.

But despite the lessening importance of smell for our species, this sense remains unique in its direct connection to areas of the brain fundamental to processing emotion – the amygdala, orbitofrontal cortex and other areas of the limbic system. And while we are less aware of our olfactory abilities than we are of other senses, they are nevertheless there. Recent experiments have shown that we humans, too, are able to unconsciously recognise body odours from the sick, to find them repulsive. It is not just danger in the form of predators or illness that we are able to detect via smell. These pheromones, or chemosignals, may actually be able to communicate emotion from one individual to another; not just negative emotions, such as fear, stress or anxiety, but also positive emotions – happiness, relaxation or sexual arousal. These studies tend to involve collecting the sweat or tears of 'donors' subjected to stimuli inducing an emotional response, such as particular videos or extreme-sports experiences. Countless studies like this have now been done, and there are some general findings. Firstly, women are better at interpreting emotions from these chemosignals than men are, but we are all generally better at interpreting these signals from members of the opposite sex than our own. The presence of chemosignals signalling negative emotions result in changes in the 'smeller', inducing defensive patterns of behaviour, modifying appetite for risk and influencing ability in cognition and perception. Positive emotions are also clearly transmitted, although it seems that these are more reliant on vision and smell in conjunction than negative emotions are. The role of these signals in attraction and mating is also apparent. Men smelling the tears of sad women compared to control tears show lower levels of arousal and testosterone.

Intimacy between partners enhances the ability to detect these emotional cues. Indeed, these chemical signals, these odours that we are perhaps unaware of, guide us in the selection of our partners, from an evolutionary perspective our single most important act, the choice that directs the continuation of our genes. Despite the long history of humans trying to change or suppress their natural odours, it seems that body odour is one of the most significant factors in our choice of mate, even more so than physical attractiveness. Apart from these chemicals allowing us to infer the health of others, it appears that they also tell us something about our potential mate's immune system. The way our bodies fight infection is through the presentation of foreign proteins to our immune cells, through a piece of molecular signalling called the major histocompatibility complex, or MHC. To have parents with different forms of MHC confirms a genetic advantage to offspring, and experiments in mice have shown that they select mates through body-odour signals that represent a different genetic immune status. Evidence suggests that this mechanism arises in humans as well, at least when procreation is the purpose. Studies have shown that women not taking hormonal contraception, or in the most fertile phase of their menstrual cycle, prefer the body odour of men with MHC genes different from their own or with a broader range of MHC genes, which confers an ability to recognise a greater variety of immune triggers. This does not appear to be the case for women on contraceptives or for men's attraction to women, implying heightened olfactory discrimination for genetically 'fit' partners when a baby may result. In fact, one study has suggested that couples struggling to conceive are more likely to be genetically similar, from an MHC perspective, reinforcing the view that procreating with someone with different genes from your own has an evolutionary advantage.

And it is not just genetic information that is conveyed by our body odour. Women in the fertile phase of their cycle prefer the sweat of men with higher testosterone levels. Scent is the secret language to safeguard your child's genetic fitness. For females, who can typically only pass on their genes to one child at a time, this is of utmost importance. But it is also the language of sexual attraction independent of procreation. Emerging evidence suggests that there are not only differences between the sexes when it comes to response to body odour, but also differences related to sexual orientation. Differences in response to single molecules in body fluids, such as male or female sex hormones, are consistently seen between both heterosexual and homosexual subjects. For example, gay men appear to exhibit higher sensitivity for androstenone – a pheromone sometimes used by farmers to encourage pigs to mate – than heterosexual men do. Even more astonishing is the finding that smell appears to convey whether someone is a potential sexual partner. One study has reported that gay men presented with body odour from heterosexual men respond differently, implying that body odour may even act as a warning in order to avoid attraction to individuals who are incompatible based upon sexual orientation.

Thus it is clear that our sense of smell extends far beyond the ability to distinguish a rose from a jasmine flower, or a lemon from an orange. Its role in more fundamental aspects of our behaviour, of our basic instincts, is clear. This sense is so ancient that it even enables communication not just within a species but between species – animals such as dogs and horses can pick up on these olfactory clues, detecting emotional state or disease in humans. Examples abound, like dogs trained to detect Parkinson's disease or Covid-19. And it all happens without awareness, a subconscious process invisible to us in our daily lives. Smell is without doubt a primaeval sense, old in evolutionary terms, primitive, and of fundamental importance to our survival.

Abi is awaiting another procedure on her nose, to see if smell can be induced. No one can tell her if this is feasible, but personally I am a little pessimistic. To regain a sense that you have lost must intuitively be easier than to gain one you never had. I am drawn to comparisons with vision. If visual problems in early life are not corrected, then the visual system does not develop properly, preventing vision from ever being fully restored. The small olfactory bulb evident on Abi's scans imply, to me, that this aspect of her nervous system has never fully developed in the absence of olfactory stimuli. Even if a good flow of air to the olfactory receptors is achieved, I am doubtful that she will ever smell in a normal way, although I would be delighted to be proven wrong.

Even if Abi's smell is regained, the past is lost for ever. Her mother, Dawn, says, 'I can go out and I can smell a smell and it takes me straight back. We're in the country, so sometimes I go past a house and I smell burning wood. My grandparents were Irish, and when I used to go over to Ireland we'd get close to their house and I could always smell their burning fire. And if I smell that smell, it takes me straight back to a place and a time. And that's one of the things for Abi – she hasn't got that. Hopefully she'll be able to make those memories but, you know, she won't smell a perfume when she's older and think, "Oh, God, that reminds me of Mum." And baking apples won't take her back to her nan's kitchen. I just like to think that some day she'll be able to make new memories.'

I consider my own smell memories. The Arabic coffee brewing in my grandparents' kitchen, the scent of a pile of damp autumn leaves and smoky garden fires, my German grandfather's eau de cologne, the flavour of my grandmother's *Pflaumenkuchen*. Another of my earliest memories is of sitting on the low wall of my kindergarten playground with my best friend. I must have been three or four at the time. Beyond the wall, the flowerbed was filled with small round pebbles of different shades of brown and cream. We would plunge our hands in, clutch several stones, and rub them vigorously against the bricks of the wall. With the pebbles warmed by friction, we would hold them under our noses, each pebble exuding a sweet earthiness, each shade a subtly different odour. Mundanity rendered memorable by smell. Memories of childhood associated with happiness or contentment, far easier to conjure than visual images or the sounds of voices, more directly linked with an emotional response. How can I say that smell and taste are the least important of the senses? I cannot. For they link our survival, our memories, our emotions, our ties to our loved ones, in the present and in the past.

7

ON THE MERRY-GO-ROUND

'We have five senses in which we glory and which we
recognise and celebrate, senses that constitute the
sensible world for us. But there are other senses –
secret senses, sixth senses, if you will – equally vital,
but unrecognised, and unlauded. These senses,
unconscious, automatic, had to be discovered.'

Oliver Sacks, *The Man Who Mistook His Wife for a Hat*

'I could hear my eyeballs moving, squashing around. As I looked left to right, I could hear the eyeball moving and squelching,' Mark tells me, as he sits in front of me. A big, jovial man, just turned fifty years old, he seems like the kind of bloke who would be centre of attention in the pub, regaling his friends with anecdotes and jokes, gesticulating wildly. But as he tells me his story, I can see him shrinking into himself, almost physically occupying less space in the room. 'It probably started seven years ago. I noticed one day that my ears felt really full, like I was in a bubble. You know when you go on an airplane and you come down and your ears are still popping? That was the sensation.' He thought nothing more of it initially, expecting the feeling to go as suddenly and mysteriously as it came. But it did not. Instead, that sensation got worse. Mark went to his doctor and was told that the problem with his ears may simply be the result of a cold, that it would settle with time.

Mark's response was typically phlegmatic – 'Fair enough!' But about a year on, he began to notice other things too. 'One thing that was really weird was that I could hear my footsteps. When your foot is going down, you're feeling it going up into your head, and it feels like a bang. The noise is echoing in your head.'

As the days and weeks passed, Mark's symptoms got gradually worse. No longer just a curiosity, the change in his hearing began to seriously affect his life. 'I quite like going out socially,' he says, confirming my first impressions of him. 'I used to go out with all my school friends on a Friday night, to the pub. But I was finding that I couldn't actually hear what they were saying. All I was picking up was loads of background noise. And if anyone put music on, my head would just be spinning.' I ask him to clarify what he means by this – is it a sensation of physical movement induced by loud noise, or is it just sensory overload? 'It was just too much. I just felt I was being hit in every direction by every sound. And I began to hear sounds that I would never have thought of before.' Like his eyeballs moving, his footsteps, his joints clicking, stomach rumbling. 'I could hear myself swallow all the time. I could hear myself breathing.' Eating, too, was a challenge. 'If I'd eat a packet of crisps, I couldn't hear anything else. Don't have a conversation with me when I'm eating a packet of crisps, because I'd ignore you. All I could hear was the crunching. I found that with toast as well.'

The fullness of the ears persisted, making Mark feel like he was in a bubble. When out and about, he would focus on the acoustics of rooms he entered. High ceilings, echoey environments, all would exacerbate the echoing of his own voice or those of others in his head. 'It felt like my head was going to explode. The amount that was going in my head was probably ten or twenty times what a normal person was hearing. Even when I was talking to people, I would lose where I was.'

Mark's change in hearing began to take its toll. Unlike the

age-related hearing loss I described in Chapter 4, Mark could hear more, rather than less. But this gain in function had a similar effect – the drowning out of what he wanted to hear pushed him towards social isolation. 'I didn't want to go out. I was making excuses about not going to the pub, because there's nothing worse than not being able to hear anyone.' I ask him if he talked to his friends and family about it, or simply swept it under the carpet. 'I've always been very positive, but I was struggling with it. I'm not sure if it is a male thing, but I didn't really want to say too much, even to my family. I made excuses – too much work, and things like that.'

Slowly but surely, various aspects of his life were stripped away. Concerts went unattended. Walking down the street was avoided for fear of hearing drivers tooting their car horns. A reluctance to walk the dog, its bark now painful to hear. 'Most disturbing were car alarms, or, where I work, fire alarms going off. People scraping their shoes, metal on metal.' And with it, his mental health suffered too. Despite knowing that this was something to do with his ears, he could not help but doubt his own sanity. 'Was it me perhaps going a bit mad?' he recalls, questioning himself.

Eventually, after several months of living in a deafening hell, Mark got to see a hearing specialist. Scans, endoscopies and hearing tests followed. 'I've done so many hearing tests! But when it is a quiet room, my hearing is really good! My tests were coming back strong. When you add something in the background, that's when my hearing totally goes.' The first diagnosis was enlarged adenoids, which were removed. The second was Eustachian tube dysfunction – an issue with the tube that connects the middle ear to the throat, which should open to equalise pressure in front of and behind the eardrum. But no treatments worked, and Mark's frustration grew. 'I probably felt the lowest I've ever felt. And I've already mentioned I'm quite a positive, upbeat person, probably the most out of all my friends. And I couldn't quite register how

I was going to have to live with this for the rest of my life.' Even at home, there was no respite; his children shouting or his wife coughing – a torture chamber.

In desperation, he sought out a second opinion, then a third. But answers were not forthcoming, and it was suggested he tried hearing aids. 'They were amazing!' he says; they had Bluetooth connectivity and were miniscule in size. But despite the technology and the high costs, they only succeeded in amplifying background noises even further. 'I tried them for a couple of months and persisted, because they're not cheap! But to no avail, and I had to take them out. The good news is I gave them to my mum, who had NHS ones,' he chuckles.

Having been on this unrewarding medical merry-go-round for a year or two now, his luck eventually changed. Having been previously told that there was nothing more to be done, a last-ditch effort was to consult a super-specialist. His heart fell when he was told that he would have more hearing tests – this would be the fourth or fifth time going through the same process. But this time the outcome was different. Rather than simply playing pure tones, there were background noises too. 'I remember one tone, a really high-pitched one – I literally fell off the chair.' He recalls the subsequent review. 'The doctor said, "We know what it is." I got very emotional, really quite tearful. She said, "It's . . ."' – at this point Mark slows down and, having stumbled over the pronunciation of it a couple of times, carefully enunciates each syllable, pausing after each word, as if stressing the importance of it in his life – '"superior canal dehiscence syndrome".'

While Mark's major problem is his hearing – making out the voices of his family and friends, being overcome by the cacophony of sounds emanating from his own body – he also reports some other rather curious symptoms. Throughout our discussion of his experiences there are a few, almost throwaway, comments that illustrate something really rather important. We've already seen that the process of turning sound into hearing is reliant on both the ear and the nervous system, and that the inner ear's function and structure are intimately interdependent. The basilar membrane, its shape, its rigidity; the size and length of the cochlear tube; the tiny series of bones in the middle ear, evolved to transmit a range of frequencies from air to fluid; the shape of the pinna, the outer ear, the whorls and ridges facilitating directional hearing – all are crucial to the conversion of vibrations in the air. But Mark reveals something else about the structure of the inner ear, something that we know theoretically but is not always apparent to us. When Mark talks about his symptoms at the height of their intensity, he tells me something very curious. At one point he says, 'That was the thing about very loud noises. They really upset my balance.' So, Mark was sent to a specialist balance clinic, where they conducted another hearing test. Mark's response to certain tones was dramatic, as we have heard, falling off his chair on that occasion. Scraping noises in particular, metal on metal, would be most certain to make him feel unbalanced.

We all know that our ears are intrinsic to our sense of balance, crucial to our knowledge of our motion through space. But this knowledge is gleaned from biology lessons, books or lectures in medical school. On a day-to-day level, however, we have no direct, tangible knowledge of this. There is no clear correlation between the organ and its balance function, much like our understanding that our thoughts originate in our brains and that our blood pressure is a function of our hearts. Our experiences tell us that our eyes see and our fingers feel – close your eyes, cover your ears, wear gloves, and those senses are modulated, dulled or eliminated. But when it comes to this relationship between balance and our ears, we are very much in the dark, like the ancient Greeks and their understanding of anatomy and physiology – they assumed that the function of the brain was to cool the heart, and that the heart controlled our thoughts, sensations and movements. However, what Mark describes is clearly a direct link between sounds and our sensation of movement, an inkling of the close structural and functional apposition of hearing and balance. For how else would we explain Mark falling off his chair in response to a particular loud sound?

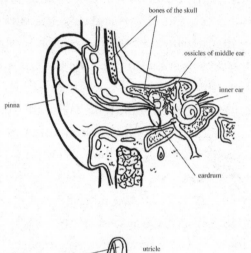

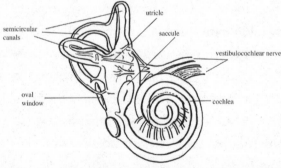

Figure 5. Above: illustration of the ear, showing the relationship between the external, middle and inner ear. Below: detailed view of the inner ear, demonstrating the close apposition of the cochlea, responsible for converting sound energy to impulses, with the vestibular system. The semicircular canals detect rotation of the head, while the utricle and saccule detect linear acceleration and head tilt.

Direct examination of the inner ear clearly demonstrates why hearing and balance should be linked. The cochlea sits deep within the temporal bone in the base of the skull, a hollowed-out curled space within the solid bone. This bony labyrinth, however, extends beyond the cochlea itself, and encompasses other fluid-filled cavities. The snail-like structure of the cochlea is contiguous with small outpouchings responsible for detecting movement of the head in space and three loop-like structures all orientated perpendicular to each other – the semicircular canals. Rotation of the head in any direction results in a shift of the fluid lining these canals, detected by hair cells similar to those in the cochlea.

Under normal circumstances, sound energy should not enter the semicircular canals and should not, therefore, precipitate a feeling of movement or rotation. But in Mark's case, something is obviously facilitating the spread of sound energy into an area of the inner ear where it should not go. The answer comes from an unlikely origin – Italian pigeons. In the late 1920s, Professor Pietro Tullio was working in the Laboratory of Experimental Physiology in Bologna. Inspired by earlier discoveries of French pigeons making sudden head movements when their inner ears were damaged surgically, he began to experiment on their Italian cousins. Opening up their semicircular canals, essentially disrupting the structural integrity of the bony labyrinth, resulted in sounds inducing abnormal eye and head movements. The parallel phenomenon, sounds inducing eye movements or a sensation of spinning in humans, was subsequently named the Tullio phenomenon. People experiencing this rare, but extremely striking condition will describe falling over when a police car drives past with its sirens on or profound dizziness triggered by a plane overhead. Originally considered to be a sign of congenital syphilis – the bacterium of the disease passed from mother to child would cause infection of the bone around the labyrinth – we now understand that this phenomenon can occur in a range of conditions in which typically, but not exclusively, a window to sound energy is opened onto the labyrinth. And indeed, the most common cause of the Tullio phenomenon is 'superior canal dehiscence syndrome' (SCDS).

I say common, but it is actually almost entirely unknown, at least among non-specialists. The condition was only described at the turn of this century, some seventy years after Tullio's experiments. In 1998, a seminal paper described a series of patients with chronic imbalance and sound- or pressure-induced spinning sensations and uncontrolled eye movements. Detailed imaging of the inner ear (the resolution of such techniques improving at around this time) revealed a thinning or breach of the bony lining of the superior canal – one of the semi-circular canals of the inner ear. Borrowing a botanical term, dehiscence – meaning the splitting-open of a seed pod – this condition was named superior canal dehiscence syndrome. The cause of this bony defect remains a mystery some twenty years later, with some researchers postulating a weakness in the bone that is present from birth while others note an association with some form of injury, like head trauma or being subjected to high pressure like the Valsalva manoeuvre – holding your nose and mouth closed against exhalation to 'pop' your ears.

Regardless of the underlying cause, this defect in the bony wall of the inner ear not only allows sound energy to stimulate movement sensors in the inner ear, but also provides a shortcut for that sound energy from outside the ear to enter into the cochlea. Under normal circumstances, as we have seen, sound energy from the air is transmitted into the cochlea through the small bones – the ossicles – of the middle ear, an efficient mechanism that also allows some regulation. The stapedius, that tiny muscle that is activated with loud noises or when we speak, dampens down transmission of some sounds. But in the presence of an alternate pathway for sound, this mechanism does not work. Sound is unfiltered, unregulated, as it passes directly from the skull into the inner ear. The middle ear – the eardrum, the ossicles and stapedius – are bypassed. Like dolphins, we end up hearing more through our jaws or our skulls than we do through our middle ear. This disrupts our hearing, but also results in 'bone hyperacusis' – hearing one's own voice very loudly and distorted, hearing your pulse or the internal sounds of your own body, such as a rumbling stomach, eye movements, breathing, footsteps or chewing. Exactly like Mark.

Mark remembers saying to his specialist that it was wonderful to know what the problem was, but asking if there was anything that could be done. 'It's very rare,' he was told. 'But the good news is that, only in the last few years, surgery for it has become possible. And there are two people in the country that can operate.' And so Mark found himself in a room with a surgeon, who asked him all the right questions. '*This is really good*, I thought. *He clearly knows what this is about.* I got quite tearful. So many years I've gone through this, and now finally someone can actually help.'

There is no non-surgical treatment for SCDS, at least not at the moment. The surgeons who originally described the condition also reported surgical procedures to plug the breach in the bone using bone dust or bone chips. The reported outcomes are good, although some patients experience hearing loss, and recurrence is possible.

Like those early patients, an operation to seal the hole in Mark's ear was proposed, and the day finally came. I ask him if he was apprehensive. 'I was full of hope. There was no worry there at all. I felt like I was in a safe pair of hands.'

Mark recalls waking up on the ward. In his excitement, the first thing he did was to call his wife and sister, but it later became clear that his conversation had been garbled and confused. 'And all this food came in afterwards,' he tells me. 'And I asked who had ordered it. "You ordered everything on the menu!" and I couldn't remember anything!' he laughs.

A day later, he was out of hospital, but the ear was swollen, his head swathed like a mummy, in huge bandages. From the first hour awake after surgery, Mark remembers constantly wondering if he could notice a difference, trying to convince himself that the operation had been a success. It was when he got home that he had the first inkling that the surgery had been worthwhile. With all the well-wishers visiting, the chatting and laughter no longer seemed intolerable. 'I wasn't getting that real buzzing in my head. I could deal with everyone talking.' The real test for Mark was about one month after surgery. 'I went to the pub, just to try. I wanted everyone there.' And as the banter started up, Mark realised that he could hear again, that he was no longer deafened by the auditory chaos of his friends. The echoing, the wall of noise, was almost gone – 'Not 100 per cent, but probably 80–90 per cent, which I would have taken any day of the week.' Sitting around a table with his friends, a pint in his hand, Mark could hear the conversation again. I ask him how he felt at that moment. 'Brilliant!' he beams. 'Absolutely brilliant! My social side came out again.' From hiding away from the world, within a few weeks he had flown off to Italy to be best man at a wedding. The flight, the speech, the party, all went without a hitch. 'I think that gave me a big lift,' he tells me.

Since that operation, some two years ago, Mark's life has returned to normal – more or less. There have been some ups and downs. Mark is waiting to go back to see the surgeon, as in the past few months, that feeling of fullness in his ears is returning. He says it's not as bad as it was, but there's a hint of deterioration. Mark is philosophical. 'I still feel lucky where I am. If we need to go back for an operation again, I would do it tomorrow.' I ask him whether his experiences have changed his understanding of life. 'I've taken for granted my hearing. But when you can't hear, you do start withdrawing from people, you do go into a depressive state. It's something that no one can see in you. They all thought, "Good old Mark, he's a laugh." But deep down I was really struggling.' Later he says, 'The thing is, I was never diagnosed with depression. It's only when I look back – that's when I see the depression. I probably put up with it in silence. I didn't share it with anyone, really. I don't think people knew how I felt at all. I was that guy that was always smiley, happy.' And as I chat to him, the confident, ebullient man sitting before me, I can still see small glimpses of the loneliness and isolation that his hearing problems brought him, and the joy of its return to something approaching normality.

We speak of the five senses, but of course this is a gross over-simplification. Other species clearly have other senses, like the lateral line of fish, a system that enables them to detect subtle perturbations of water current indicative of prey or predators, the pit organs of snakes, allowing them to detect heat radiation, or bird species able to perceive the Earth's magnetic field. In humans, proprioception, these sense of where our bodies are in space and where our body parts are in relation to each other, as was lost by Rahel due to her lung cancer (see Chapter 1), has been described by some as a sixth sense, although to my mind this is simply an unconscious modality of touch.

However you define them, it is clear that Aristotle's view

of the *five* senses – a view that we hold on to today, some two millennia later – is incorrect. Ask anyone who has ever experienced seasickness, or staggered off a rollercoaster with that profound nauseating feeling of constant movement even after you have reached terra firma. Knowing which way is up, which way is down, or perceiving how our heads are moving in space, is fundamental to our lives. It permits us to walk in a straight line, to stand against gravity. At its most basic, even being able to see clearly is dependent on this sense. Consider walking down the street, your head moving up and down, tilting forward or back, side to side. Our visual field should be jiggling all over the place, juddering furiously as we walk or run. But it doesn't; it remains firmly fixed in place, allowing us to focus on the person's face strolling down the pavement towards us. And this is all down to the inner ear, thanks to direct connections between the vestibular system and the motor supply to the muscles of the eyeballs, a reflex maintaining the eyeball in position despite the movements of the head in space. This is referred to as the doll's eye reflex, like the children's toy whose eyes remain pointing in the same direction relative to gravity, making them appear to close when you lie the doll down.

Nothing illustrates the importance of balance more than when it goes wrong. My shoes bear testimony to that. One of the most common balance conditions we see in the neurology clinic is benign paroxysmal positional vertigo, or BPPV. Vertigo in the medical sense defines the sensation of movement when there is none, very different to fear of heights. Individuals with BPPV will often experience a profound sense of dizziness or spinning triggered by small head movements or rolling over in bed. The cause is something overtly insignificant – small solid particles in the fluid that fills the semicircular canals of the inner ear. When the head moves, the fluid inside the ear moves, but as the head stops, the fluid no longer moves. However, any solid particles will continue to drift, giving the illusion of movement, which is extremely disconcerting and unpleasant. In some respects, someone with BPPV is a godsend in the midst of a busy clinic – a simple manoeuvre on the examination couch can usually diagnose and treat the condition in five minutes flat. There are very few instant hits in the world of neurology, but this is one of them. The downside to this manoeuvre is that it precipitates the very spinning that has brought the person into clinic in the first place; I have learned not to let the patient grip my arm, after the first few times of having nail imprints on my wrist as the patient clutches on for dear life while the world spins around them. And after the first time of having my shoes covered in vomit, I have also learned to stand well back and keep a bowl to hand.

Sometimes, however, vertigo can strike even without a trigger such as turning your head or looking up. And Kelly knows exactly how devastating that is. Sitting in front of me, she looks healthy, full of life, relaxed and confident. In her early thirties, she looks younger, with a long mane of blond hair and fashionable thick-rimmed glasses. Like Mark, she laughs easily, even though her condition is no laughing matter.

'It all started in 2014. I was sat at my desk at work. I was in recruitment, so it's really fast-paced. And I was staring at my screen and everything just started spinning. So I'd no idea what was going on. And then it stopped. And I thought nothing of it and just carried on, as you do,' she chuckles. She remembers the episode lasting a second, over so quickly it was difficult to know what to make of it. She thought she had just not drunk enough water, or hadn't eaten enough, and just carried on. Over the next few weeks, however, it happened a few more times. 'The world was spinning round me,' she recalls. 'I was hanging on to the chair for dear life!' And these attacks began to happen while she was standing up or walking down the road. 'It would start with really loud tinnitus in my left ear, like a very loud constant beeping.' She mimics a high-pitched tone, prolonged and grating. 'And then my ears would fill up – like when you are on a flight and you can't hear – and then I would fall straight to the floor. It was very scary.' I ask her if at that point she had any idea what was going on. 'No, not a clue. I went to the GP and they gave me some tablets for dizziness. They thought I had labyrinthitis.'

Labyrinthitis is essentially the default diagnosis for anyone with dizziness. It is a catch-all term, used and abused, but is thought to represent infection or inflammation of the labyrinth, the bony canals of the inner ear, causing disruption to hearing and balance, and resulting in nausea and vomiting. Often presumed to be due to a viral infection, the term is often used interchangeably with vestibular neuronitis, inflammation of the nerve to the labyrinth, and manifests in the same way. While no one could accuse Kelly's condition of not being dramatic, labyrinthitis is even more so, with the afflicted suffering profound constant spinning, extreme vomiting and being unable to get out of bed for several days before their condition gradually improves. Not at all what Kelly was experiencing. 'So, I took those tablets for a few weeks. They didn't work. I'd walk out of my house and fall to the floor. I remember chatting to my neighbour and having to try and wobble my way back to my house without falling over in the street. I literally fell through my front door.'

As time passed, Kelly's episodes grew longer and more frequent. Three or four times per week, for five or six hours at a time. There was nothing to do but wait it out. 'I would be sat on the floor, being sick. Once everything started spinning, there would be nothing you could do. Strangely, being sick would give me a few seconds of relief.' Her symptoms quickly began to dominate her life. The tinnitus started to occur between these attacks as well, and after the attacks she would lose some of her hearing for a short while. If she used her phone on the left ear it would be very muffled, yet it remained crystal clear on the right.

Kelly returned to her doctor, who then thought she might have an ear infection. She was prescribed antibiotics, again to no avail. 'I remember at this point it must have been my fifth GP visit. I was in tears, saying there is something wrong, and whatever you're telling me isn't right.' A third diagnosis was forthcoming – Ménière's disease. Kelly recalls, 'They told me to take these tablets – betahistine – and see if it worked. They then sent me home again.' I can hear the resentment in her voice, the clear feeling that these horrific-sounding symptoms were not being taken seriously, that she herself was not being taken seriously.

I ask her if the new medication worked. She laughs. 'No, unfortunately!' I tell her I imagine she was not laughing at the time. 'I was not! I laugh now because I realize that I'm not the only person that has been through this. You hear me laughing a lot, but that's the only way you can get through it. Otherwise you just end up sad and upset all the time. At that moment, you feel very alone, very isolated. I was twenty-seven; I had never had anything like this. I've never really had vertigo, or dizziness. You know when you were a child and you went to a park, and you all got onto one of those merry-go-rounds, and your friends would spin you and spin you and spin you. And then you would get off it and you'd be walking – you can't walk straight, and you end up falling on the floor because everything is spinning? But that stops within a couple of minutes, even less than that. Not this – this carries on and is ten times worse, and it can last hours. That feeling of no control of your body, basically.'

Five months on, Kelly was back at her doctor's. She had not been to work for most of that time. An attack on the Underground had left her stuck on the Tube platform for three hours. Kelly had got off one Tube train and was walking to another line. 'I started having to hang on to the walls as I walked,' she tells me. 'I didn't make it onto the next train. I made it to the platform and collapsed on the floor.' A passer-by, who Kelly assumes to have been a doctor, stopped and called for an ambulance; Kelly recalls lots of people walking by, assuming she was drunk, despite it being 8.30 a.m. Help in the form of an ambulance crew was unfortunately not forthcoming, as Kelly was not deemed an emergency. She remembers eventually having been taken to hospital by a British Transport police officer in a taxi.

It was after this incident that her GP finally referred her to a specialist. But Kelly snapped when she found out she would have to wait three months for this appointment. 'After a week, I called up the hospital, spoke to the Ear, Nose and Throat secretaries, and basically burst out crying, telling them I had no life. I was stuck at home. I couldn't do anything. I was completely isolated. I had no independence. At this point, I knew I was going to lose my job. I needed to see someone and find out what was going on. And the lady I spoke to was absolutely incredible. She got me an appointment the next day. It gives me goosebumps thinking about it.'

The following day, sitting in the clinic, Kelly got some unexpected news. The GP had been correct in the final diagnosis offered, despite the lack of response to medications. 'The consultant said that I had typical symptoms of Ménière's disease.'

The Paris of the mid-nineteenth century was a hotbed of advancement in the medical sciences. In one corner of the city, at the quackery end of the spectrum, the followers of Franz Mesmer, the German physician who espoused animal magnetism and had passed briefly through Paris, practised mesmerism. But elsewhere in the city, huge strides were being made in crafting a scientific method to the practice of medicine. An innovative group of physicians had developed a framework for teaching and research at the turn of the century, setting the stage for the advancements to come. War with the British, and resultant blockades preventing the importation of substances from the tropics, kickstarted the local pharmaceutical industry and had led to the discovery of strychnine, quinine, caffeine and codeine. René Laennec had been lauded for the invention of the stethoscope in 1816, and the introduction of auscultation – listening through his device – into clinical practice. By 1857, Louis Pasteur had moved to the city from Lille and had already begun to work on fermentation and the germ theory of disease. In the world of neurology, meanwhile, Jean-Martin Charcot, one of the founding fathers of the modern speciality, had begun his career as a physician and was soon to be appointed a professor at the University of Paris. It was he who laid the foundations for neurological clinical examination, describing countless neurological disorders, including multiple sclerosis and motor neurone disease. The list of Charcot's acolytes is a Who's Who of neurology, names instantly recognisable to most doctors.

It was into this brave new medical world, in 1861, that Prosper Ménière took his first steps, at the tender age of sixty-one. To describe him as a wide-eyed ingenue is perhaps a little unfair. By this time, Ménière was already a respected physician in Paris, the director of the Imperial Institute for Deaf-Mutes, but in some respects he was not an accepted member of the medical establishment. On two separate occasions he had tried to join the Imperial Academy of Medicine, but had failed to muster sufficient votes for election.

It was in January of that year that Ménière stepped behind the podium at the academy that had thus far shunned him, to present a paper. By some accounts, it was a miserable day outside, bitterly cold and raining hard, and few had turned up to attend his lecture. The scant audience that had appeared showed little interest in what he had to say, and, to add insult to injury, Ménière was not permitted to engage in the discussion afterwards, having failed to be elected a member of the institution.

At the time, vertigo – the sensation of spinning or movement – was thought to be brain-related. Termed 'apoplectic cerebral congestion', it was presumed, like strokes and seizures, to be caused by the overfilling of the blood vessels in the brain and was treated by bleeding, cupping, purging or the application of leeches. Other gruesome treatments in use at the time included setons – large, curved needles threaded with string, left in the neck. After a few days, the string would cause suppuration (the formation of pus), which was considered to draw out inflammation.

In his presentation of his paper, Ménière controversially proposed an alternative explanation for vertigo. Referring to his patients at the Institute for Deaf-Mutes, he described those who had developed sudden deafness and vertigo after a foreign object had been inserted into the ear, and others in whom hearing loss and tinnitus developed hand in hand with vertigo. He concluded that vertigo may actually originate from the ear rather than the brain. Perhaps most convincing was his description of a young girl who had presented with extremely sudden vertigo and hearing loss, and died a short while later. At the autopsy, her brain and spinal cord were found to be intact, but within the inner ear was evidence of bleeding, with blood filling the semi-circular canals rather than the cochlea. This unfortunate child – in hindsight probably suffering from acute leukaemia, leaving her at risk of spontaneous haemorrhage – constituted the most direct evidence to refute previous views of the origins of vertigo.

On the day of his presentation, Ménière's views were largely ignored, but they were resurrected the following week by a figure more closely associated with the establishment, and a member of the Academy, who cited his paper while talking about epilepsy. The resulting controversy raged for weeks, until the academician had had enough and withdrew from the debate. Disenchanted, Ménière nonetheless persisted in his collection of cases, and published regularly until his death of influenza in February 1862, some thirteen months after his controversial paper.

Although dismissed and rejected by his contemporaries, Ménière's name lives on ever more in his eponymous disease. While we now clearly understand that vertigo may originate from the ear (it can of course also result from the brain, in the cases of migraine or stroke, for example), there remains some mystery regarding the underlying cause of Ménière's disease. The condition manifests just as Kelly describes, with sponta-neous episodes of spinning lasting from a few minutes up to twenty-four hours, as well as fluctuating hearing loss, ringing in the ears and a feeling of pressure in the ear. The disease is often correlated with the accumulation of endolymph – the fluid filling the cochlea and vestibular system in the inner ear – but the relationship between this fluid build-up and the disease is not fully understood: people can develop this same build-up of endolymph without any symptoms at all. And the reason for this accumulation of fluid in the first place is unclear, although it is presumably related to either too much production of the fluid or too little absorption. Ménière's disease is considered a disorder of middle age, but can occur in children as young as four years old and in the elderly. Genetic, cellular and anatomical factors are all thought to contribute.

Why this build-up of fluid, termed 'endolymphatic hydrops', should result in such symptoms is only a little better understood. Accumulation of fluid causes the basilar membrane in the cochlea – the location of the hair cells – to bulge, causing changes to sound transmission and detection within the inner ear, and resultant hearing loss. But the reason for the tinnitus and vertigo is more uncertain. Some studies suggest that Ménière's disease also results in damage to the nerve fibres conducting information from the inner ear. It has also been suggested that rupture of the membranes of the inner ear can lead to leakage of endolymph, whose high potassium levels can cause nerve fibres to fire spontaneously.

For Kelly, and for others with the condition, one of the most disabling features is the presence of drop attacks or 'Tumarkin otolithic crises'. Kelly describes scenarios where, without warning, she finds herself on the floor. There is no loss of consciousness, but equally no vertigo or imbalance beforehand. It is as if she is struck to the ground by an invisible hand, out of the blue, one moment walking along the street, the next moment spread-eagled on the paving stones. So striking are these collapses that I have seen several people with Ménière's mistakenly referred to my epilepsy clinic, their doctors at a loss to explain why this person with intermittent vertigo may suddenly drop to the floor. The cause is thought to result from a sudden dysfunction of the gravity-sensing organs of the inner ear.

Kelly was put back on betahistine tablets. She recalls that the surgeon also proposed some intra-tympanic steroid injections – passing a hypodermic needle through the eardrum, injecting steroid directly into the middle ear. 'It hurt a lot,' Kelly chuckles. 'It wasn't so much the needle going through, but the liquid inside. Your ear feels raw. It was the most painful thing I've ever felt.' I almost feel uncomfortable doing so, as I suspect I know the answer, but I ask if these injections helped. 'No,' she laughs. 'I had three sets of them every two weeks, but they didn't really help.' Throughout this time, even with a firm diagnosis, her life had come to a standstill. Her mother, living abroad, was flying back and forth to look after her. 'I wasn't allowed to go out on my own; I had to have someone with me the whole time. I wasn't allowed to go to the toilet and lock the door; if something happened in the toilet, no one would be able to get in there and help me. Basically, I was twenty-seven with no life.' Over a period of a few months, Kelly had gone from having a normal, active lifestyle to being housebound.

After eighteen months of suffering with the condition, she lost her job. She admits it was a bit of a relief. 'It actually helped me, because I didn't have the stress of constantly having to say that I can't come in this month.' She was fitted with a grommet – a small plastic tube to maintain a hole in the eardrum, allowing air to pass in and out of the middle ear – in an attempt to help. Whether by coincidence or due to the surgery, it was shortly after the procedure that Kelly began to see some signs of recovery. Kelly went back to work but, recognising that stress was a significant trigger for her, in a limited and junior capacity. Gradually, some semblance of normality returned, but while she still gets dizzy spells – 'If I'm stressed, I'll be on the floor' – they are rarer now. She still gets the tinnitus, too, but intermittently rather than continuously. 'I've lost a lot of hearing in my left ear, so I do have a hearing aid now – when I remember to wear it!' she scoffs. 'When it's really busy around me, if I put it in I can actually hear someone talking to me, whereas otherwise I'm lip-reading. And according to my parents, if I have the hearing aid in I speak at a normal level. I don't shout.'

Kelly works part-time now, and was mainly working from home even before the Covid-19 crisis. She has also moved out of London, where it is quieter. The lack of hustle and bustle has made life easier. I ask her if she still has a nightlife. 'I still go out, but I don't drink. The last thing I want to feel is drunk, when I feel that normally anyway!'

What is clear is that there remain as many unknowns about the treatment of Ménière's disease as there are about the cause. Migraine often complicates Ménière's disease, as it does in Kelly's case. Management of stress levels, and other lifestyle factors such as sleep and alcohol intake, may well help both migraine and the symptoms of Ménière's. Specific treatments are more hit-and-miss. Betahistine, the drug prescribed for Kelly, suppresses an overactive vestibular system and may improve blood flow to the cochlea, but ultimately the precise way it works is unknown. The inner ear can absorb substances from the middle ear, hence the injections of steroids through Kelly's eardrum. These procedures may reduce the number of spells of vertigo but do not eliminate them. More aggressive treatments have been tried, such as decompression of the inner ear, the insertion of shunts to drain endolymph, or even complete destruction of the balance-sensing organs of the inner ear or the nerve, with variable outcomes. There remains uncertainty as to whether some of these treatments ultimately alter the natural progression of the disease. The insertion of a grommet into Kelly's ear appeared to help, but is certainly not universally accepted as a treatment for Ménière's. Some surgeons have noted that grommet insertion may abolish attacks by causing changes in pressure in the middle ear, and ultimately in the inner ear too.

I ask Kelly if she thinks her improvement is down to anything that the doctors have done – the tablets, the injections, the grommet – or whether it is simply due to the condition moving into a better phase. She shrugs. 'Unfortunately, I don't think even the doctors know enough about it. I think it has gone into remission, and some of it is down to me trying to get on with it. I don't have another choice. My choice is either stay at home and be anxious about going out, or worrying about having an attack when I'm going out.' Kelly has clearly taken a decision about how she wants to lead her life. 'If it happens, it happens. Sometimes that anxiety of worrying about the Ménière's can actually bring on an attack. That's something I noticed a lot. If I left my house and I worried about it, it happened.'

For the moment, at least, Ménière's has weakened its grip over Kelly, physically and psychologically, but whether that's through its natural progress or down to the various medications and procedures remains to be seen. For the moment, Kelly is simply thankful to have regained some of her life, but she remains uncertain as to what the future holds. Is the Ménière's disease just quiescent, soon to resurface, or has it been beaten into submission?

As I listen to Mark's and Kelly's stories, they represent yet another reminder of the vulnerabilities of our senses, and indeed our lives. A tiny thinning of bone, or a slight over-production of fluid, can render us unable to hear, unable to socialise, incapable of walking in a straight line, impotent in the face of gravity. Tiny defects in our anatomy, in some cases so miniscule that they have not been noted until a few years ago, can influence our perception of sounds, of posture, of up and down, and of movement – all in life-changing ways. And these stories are another example of the fundamental role our senses have in our relationship with the world outside us, the physical, psychological and social environment we inhabit – even those senses that we do not recognise as such, that we are not even aware of. Once again, they form an aspect of our life that we notice only through its absence.

8

THE BURNING TRACKS
OF MY TEARS

'It is strange that the tactile sense, which is so
infinitely less precious to men than sight, becomes
at critical moments our main, if not only, handle
to reality.'

Vladimir Nabokov, *Lolita*

'I sometimes wonder if the hand is not more
sensitive to the beauties of sculpture than the eye. I
should think the wonderful rhythmical flow of lines
and curves could be more subtly felt than seen. Be
this as it may, I know I can feel the heart-throbs
of the ancient Greeks in their marble gods and
goddesses.'

Helen Keller, *The Story of My Life*

After inhabiting the same clinic room in my hospital's neurology department for over a decade, I am still unable to master
the temperature control. In the height of summer, the air conditioning keeps me comfortable, although there is a faint whiff
of sewage, or perhaps a dead rat, emanating from the grates
in the ceiling. In winter, however, the temperamental heating

means the room is decidedly frosty. There is usually a race to find one of the few plug-in radiators in the department, but if my colleagues are ahead of me and grab them first, I will shiver my way through my morning clinic. I rarely miss the days when we all wore white coats, but at times like this I would be grateful for an extra layer to hold the cold at bay.

It is on one such a winter's day that I meet Miriam (not her real name). London is in the midst of a very cold spell, and occasional flakes of snow fly past the window, interrupting my view of the Shard. I can see a tide of people pouring out of the Tube station, zipping their coats under their chins and turning up their collars as they face the icy wind. But as I call Miriam in to the clinic room, I am struck by her odd attire. In her late forties, with short dark hair, she is wearing a thick down jacket and holds her gloves and scarf in one hand, but as I look down to her feet I note, to my surprise, that she is wearing sandals. Between the straps, the skin of her feet glows bright red, in stark contrast to my white fingers, slowly defrosting from my journey into work. Her teenage son silently accompanies her, with footwear more appropriate for the weather outside. Once they are both settled, we begin to explore her problems.

'My feet have burned for as long as I can remember,' she tells me. 'Even as a child, I would hate wearing shoes. No matter what temperature it was outside, they would always feel hot.' She describes periods lasting hours or days when her feet would feel incandescent, as if thrust into a fire, often triggered by exercise, but also just by sitting; sometimes even without a clear cause. But over the years, this burning has become much worse, to the point that it is positively agonising. Now, even when her feet should be icy cold, the burning continues. Even in the chilly clinic room, she shifts her feet uncomfortably and occasionally winces as she talks. 'The rest of me feels cold but my feet are just on fire. They feel like they are in a furnace.' Sleeping is an issue: 'I constantly seek out cold spots in the bed and cannot bear to have the duvet over my feet. It is exhausting. As soon as I start to feel sleepy, the burning means I have to move them to find the next cold part of the bed.' She has begun to put her socks in the freezer, wearing them as she gets into bed, or using icepacks to obtain relief. 'I have noticed lately that I am actually getting frostbite,' she says. The treatment for her discomfort is damaging her toes.

When I examine her, I look carefully at the skin of her feet. Apart from the puce hue, slowly fading to normal somewhere just above her ankles, they look rather unremarkable. This sort of burning sensation is something I see most frequently in people with severe diabetes, where the persistently high levels of glucose damage the blood vessels that supply the nerve fibres, the conduits for sensory impulses from the skin to the central nervous system. Starved of an adequate blood supply, the health of the nerve fibres themselves then suffer. But with diabetes, I would expect to see some other evidence of nerve damage – some numbness, perhaps some weakness, or at least a change in reflexes – yet with Miriam there really is nothing.

The devastation from the loss of touch-sensation is readily apparent. Take Paul's loss of pain through a genetic mutation rendering him destined to damage himself beyond repair; the absence of proprioception leaving Rahel immobile and infirm. But the opposite can be equally destructive. Too much sensation – the perception of it in its absence, or the amplification of it – can also have life-changing consequences. Think about the unpleasantness of waking up having slept on your arm, the painful tingling as the circulation is restored; or when you have been sitting cross-legged for a little too long – you go to stand up, and the pins and needles, and numbness of your foot, cause you to almost stumble with discomfort and the inability to feel your foot on the floor. Now multiply these feelings a hundred- or a thousand-fold; imagine that these sensations, now infinitely more intense, were also no longer transient but permeated every aspect of your waking life. As with loss of sensation, sometimes the cause is damage to our nervous system, occasionally to a function of our genes. And as with Paul and Rahel, the effects are surprising, shocking and transformative, but also tell us much about how we all make sense of our skin and what it tells us of the world within and beyond us.

As well as being a barrier between us and the outside world, the skin is also a bridge between exterior and interior, a conduit for moisture, temperature and chemicals – but clearly also sensation. An array of sensory organs are embedded in the tips of our fingers and the ends of our toes, indeed throughout our skin – organs to detect pressure, delicate changes in texture, skin movement; others to detect light-touch or the sense of vibration. Some are specialised to detect the bending of a hair in its follicle. These tiny organs, like pressure pads on the floor of a bank vault, sit on the endings of nerve fibres, with each individual fibre carrying only one specific type of receptor, so that one fibre may only detect skin movement, another only texture. But this subdivision of labour does not stop here. Fibres of different size or shape each conduct different sensation types, like telephone cables each carrying a different conversation. The smallest fibres are those largely responsible for pain and temperature, and seem particularly prone to damage by diabetes and other specific diseases. Unlike other sensation types, there are few specific organs in the skin dedicated to detecting pain. Instead, the endings of these nerve fibres float free, and are themselves the detectors of pain.

But how do these nerve endings actually detect the presence of something that might cause the body harm? Over recent decades, numerous different molecular receptors have been identified on the tips of these nerve fibres, each detecting different triggers, such as heat, cold, inflammation or acid. In the presence of these triggers, the nerve endings generate signals that are propagated up the nerve fibres, the very first step from periphery to brain, the origin of sensory perception.

Even at this molecular level, perception and reality diverge. In the past few years, specific receptors have been identified that detect not just temperature, but very specific temperatures. For example, one specific detector will be triggered only by temperatures above 42°C; another at less than 17°C. But it seems that hot and cold are not the only triggers for these receptors. Through the power of evolution, Nature has worked out how to trick us. Plants have developed substances that can also make us perceive hot or cold, be it to attract us or repel us from touching or eating them. Remember your last outing to a Thai or Vietnamese restaurant, that gentle warmth in the mouth as you eat your meal. As you bite down on a wayward sliver of chilli, that warmth is replaced by a jolt of burning as your mouth is flooded by a compound called capsaicin, released from the chilli. It turns out that capsaicin is a natural trigger of one of these receptors, tricking the nervous system into experiencing a sensation of damaging heat. Similarly, the 'minty cool' that is a favoured slogan of advertisements for chewing gum or toothpaste has its basis in this trick of Nature. Menthol or eucalyptol, chemicals in mint and eucalyptus respectively, activate receptors that detect a drop in temperature, giving us that feeling of coolness in the mouth or the skin. And we now know that other plants, too, like garlic, cinnamon and horseradish, contain substances that seek to confuse us, to convince us that we are harming ourselves when we eat them. But Nature's trickery has backfired. We have learned that these substances do not harm us, and now actively seek out these plants for their sensations and tastes, demonstrating the fine line between pain and pleasure. The original aim, to stop us eating these plants, has actively led us to seek them out. But perhaps this explains why children tend to avoid these foodstuffs. They have not yet learned to appreciate the joy of garlic-laden pasta or a spicy pad thai.

If you damage your small nerve fibres, your ability to detect the pain of a pinprick may be either altered or lessened, and although you may experience pain, your ability to detect temperature would also be impaired. Conditions like leprosy cause the destruction of nerves that result in an inability to feel pain or temperature in the extremities of the body. Diabetes may cause numbness, but it may also cause irritation of these nerves, resulting in debilitating persistent pain. For Miriam, though, this is not the case. Her nerves are entirely intact.

On a hunch, I ask her if anyone else in the family has a similar condition. She laughs, as does her son, who has remained absolutely silent until now. To my surprise, he begins to remove his white trainers and sport socks. 'My father has it, and so does my son,' Miriam says, as he peels back his socks to reveal feet the same violaceous colour as hers. 'I always thought it was just one of those things. I have had it almost all my life, and saw it in my father when I was growing up.' It is only the worsening agony of burning that has led her to seek medical help; otherwise she would have simply continued to live with it. Her son is also rather nonchalant about his feet, as he has never known any different. I ask him if he has the same burning. 'Yes, a little, but it does not bother me too much, especially on cold days like this. In the summer, though, I can't wear shoes.'

As her son exposes his feet too, I feel a slight tingle of excitement that I may have made a diagnosis – a condition once read about but never seen.

In medicine, there is an old adage of 'When you hear hoof beats, think of horses, not zebras,' ascribed to Dr Theodore Woodward, a Maryland professor of medicine in the 1940s, his point being that one should look for the most likely explanation for a set of symptoms and signs, not look to a rare and exotic diagnosis. As neurologists, we pride ourselves on an encyclopaedic knowledge of strange and rare conditions (nowadays assisted by the internet), a throwback to the days when we had little to offer except a diagnosis itself. Thankfully those days are long gone, and we have an arsenal of treatments to offer, but this spirit of seeking out the rare, those cases that have gone undiagnosed before, continues to be woven through the fabric of our speciality. We like to consider ourselves as hunters of the zebras (or even better, unicorns), seeking out the rare beast among the vast herd of horses. And as I gaze at the two pairs of glowing feet in front of me, belonging to Miriam and her son, I think I recognise one of those zebras.

'I'm probably not like your normal hotel guests, who eats a varied diet. I ate the same sort of thing every evening for dinner, which was just a locally caught coral trout. Beautiful meal: organic, fresh. And I would eat that every night for probably the ten to fourteen days that we were visiting. It was delicious.' I speak to Alison down the line, she in her hometown of Sydney, me in London. She is telling me about her family's favourite destination for rest and relaxation, a resort in Fiji so beautiful and tranquil that they would travel to the same hotel year after year. It sounds idyllic; I have visions of palm trees swaying in the breeze, white-sand beaches and crystal-clear turquoise sea, and am overcome by a pang of envy as I look out of the window at the steel-grey London sky. As she describes her fantastic meals, I gaze at my lunch of a cheese sandwich, feeling short-changed. I have never met her face to face, but googling her brings up a picture of a slim, polished blond woman, perhaps in her late forties, and in my mind's eye I see her, her husband and their three children in that five-star resort, away from the stresses and strains of normal life back home.

However, in 2013, several days into one of their regular trips, Alison experienced something odd; something felt not quite right with her feet. 'I was washing my hands, and I had a strange sensation: when I walked across the tiles, it felt like my feet might be a little bit sensitive. But it came and went, and it was fairly mild, so I didn't really think about it too much.' On her return to Sydney, she mentioned it to her doctor, but as these strange sensations had disappeared, they were dismissed.

It was not until the following year, back in Fiji, that she noticed something again. But this time is was not just her feet. 'When I was washing my hands, it felt like hot was cold and cold was hot. And after about twenty-four to forty-eight hours, I was eventually finding it hard to even hold a glass of cold water. It felt like my fingers were burning around the glass.' Her sensation of temperature in her hands had been switched around, inverted to the point that what should have felt like refreshing icy-cold water felt scalding hot, and the warm water of the shower felt shockingly cold. The rules of the world as we know them had turned on their head.

But the reversed sensation was not just in her hands. 'If I put the water to my lips, it felt like my lips were burning.'

As she walked across the cool tiles of the bathroom floor, the soles of her feet burned. 'If I tried to step into the pool, it felt like my feet were burning in the water. And then the pain was so bad that it would make me cry. And the tears felt like they were burning my face because it was that sensitive. It was very frightening. I didn't know where it was going to end.' I ask her if there was anything to see – a rash, a redness or any external manifestation of illness or dysfunction. Alison says, 'I mean, the family photos are fantastic. I looked the epitome of health. I had a tan, I was well, no fever, no vomiting, no diarrhoea. But I'm actually really in pain the whole time.'

With Alison in a world of pain and confusion, her grip of physical reality waning, she and her family were terrified. They called on some medical connections they had and after a little while managed to speak to an experienced doctor, who suggested a possible explanation. The answer is a surprising one, and explains both Alison's burning tears and Miriam's glowing toes. In fact, it is an entity that you will already be familiar with: the sodium channel – and specifically, just like the cause of Paul's congenital insensitivity to pain, the Nav 1.7 sodium channel. In some respects, Paul is the very mirror image of Alison and Miriam: on the one side, a total absence of pain even with intense provocation, on the other, intense pain without a cause.

For Alison, the diagnosis is in her nightly fish supper. Everyone is familiar with the dangers of seafood on holiday – a strange-tasting fish stew, some dodgy mussels, and the resultant upset stomach. But to have such symptoms as Alison's, with a reversal of temperature-sensation as a result of food poisoning, seems bizarre. However, it is not the fish itself that is causing the poisoning; it is what the fish has been eating. The poison comes from lower down the food chain.

Ciguatera poisoning is almost unheard of in the UK, but will be more familiar to neurologists working in the Caribbean or around the Pacific, and globally it is thought to be the commonest form of fish poisoning, with up to 50,000 cases worldwide every year. The toxin, ciguatoxin, is the product of a plankton that is found on coral reefs and ingested by certain fish, which are then eaten by other, larger fish, the toxin then gradually concentrating in their tissues, especially the head, roe and skin. The bigger the fish and the higher up the food chain, the more ciguatoxin it is likely to contain. Unlike many types of food poisoning, ciguatoxin is not destroyed by cooking or freezing, and it is not caused by the mishandling of food. Odourless and undetectable, the toxin does its work when we eat contaminated fish. Once ingested, it is rapidly distributed throughout our bloodstream and around our bodies, binding to targets in our gut, heart, brain and nerves. Disruption of normal function in these organs ensues, sometimes resulting in diarrhoea, vomiting and abdominal pain, changes in heart rate and blood pressure, and difficulties with concentration and memory. But it is the damage caused to the peripheral nerves that causes ciguatera poisoning's most striking symptom, that of 'hot–cold reversal'. While the precise mechanism by which this happens remains obscure, an increasingly familiar culprit is to blame. Ciguatoxin binds to the ubiquitous Nav 1.7 sodium channels, causing their dysfunction and, consequently, abnormalities in the operation of the small nerve fibres acting as the conduit for pain and temperature information. Confusion in these pathways causes bewilderment of our nervous systems and the cold tiled floor of the hotel room is perceived as the burning embers of a fire walk. But while the nervous system tells a lie, reality remains constant – a block of ice held against the skin may feel burning hot, but will still cause frostbite.

In truth, I have always been fascinated by ciguatera toxicity, since the very first time I read about it. In neurology, we are used to people experiencing things that aren't there or not perceiving things that are there. But this is different: a single molecule, derived from plankton, scrambling our senses to such a degree that the rules of physics appear to be upended; that hot is cold and cold is hot, like Satan whispering in your ear, making you believe the opposite of what is true. It is perhaps one of the best examples of how, when it comes to your understanding of the world around you, you can be deceived or betrayed by your own body, your own nervous system.

Alison was the only member of her family to be poisoned like this, however, and no one else in the hotel was afflicted. The answer may lie in her lack of dietary variety: the same dish, specifically coral trout, night after night after night – 'delicious, organic, freshly caught' but laden with ciguatoxin, the poison trickling into her with every meal, building up with each idyllic day in the sun.

On hearing a possible explanation of her symptoms, Alison returned to Sydney, where a diagnosis of ciguatoxin poisoning was confirmed. She was put on medication that eased the discomfort, but the painful temperature inversion persisted for three to six months. She continued to have flare-ups for a couple of years afterwards, triggered by periods of illness or stress. Even now, some six years later, Alison still lives in the shadow of the coral trout, occasionally experiencing symptoms when sufficiently stressed. She tells me, 'Especially in the first stages of recovery, I took no medications, no caffeine, that would stimulate it; if I drank caffeine, I'd have burning around my mouth. I've also avoided alcohol for the past six years. I've tried it on and off, and found that it made me feel quite strange again.' I ask Alison if she would ever eat fish again. Unsurprisingly, she says,

'For me, the experience was so severe that it's taken a lot of years to recover. I think that I will probably never eat seafood again.'

Before anyone who does not live or holiday in these exotic locations sits too comfortably, they should heed a word of warning from Matthew Kiernan, Alison's neurologist in Sydney, who believes ciguatoxin may be affecting people without them being aware of what is causing their symptoms.

Fish that are caught in the Caribbean can easily be on a restaurant table in London that evening. Similarly, fish throughout Asia–Pacific are distributed through very efficient networks. I think that there's a lot of exposure going on, but probably not so much realisation. So every time someone gets what would be called food poisoning after going and eating fish at a restaurant, there's no understanding that it might be ciguatera related to reef fish. And when you order reef fish in a restaurant in London, you don't ask, 'How big was the fish? Where was it caught?' So if a restaurant offers coral trout, it seems pretty exotic and a lot of people would go for it.

Miriam and her son, however, have never been to Fiji, and the answer for them is less exotic. As the family history would suggest, the cause of their problems is not poisoning, or indeed any other external factor. It is in their genes. And when I say they are the mirror image of Paul, this is almost literally the case. For while Paul has mutations in the SCN9A gene that abolish production of the Nav 1.7 sodium channel, the molecular apparatus responsible for conduction of pain, Miriam has the opposite. Her family has a mutation in SCN9A that results in overactivity of this sodium channel. She and her son have a condition called primary erythromelalgia, or PE.

Since the original description of a mutation in this gene, in 2004, more than twenty different mutations at different points in the gene have been identified in families with PE. But they all have similar consequences. Each mutation results in subtle changes to the structure and function of the sodium channel, altering its properties, how easily it opens and how long for. And even such minor changes as this result in the nerve-fibre endings being triggered under normal conditions, rather than in response to scalding water or a raging fire, creating the illusion of painful burning when there is none. But these types of sodium channels exist not only in those pain-sensory fibres. They are also present, in large concentrations, in other nerve fibres weaving through the skin – so-called 'sympathetic' fibres, part of the nervous system that is largely hidden from our own awareness, regulating bodily functions like heart function, blood pressure and movement of the gut. This network of nerves also regulates the constriction and dilation of blood vessels, and its dysfunction in PE causes small capillaries in the skin to engorge – thus the puce colour of Miriam's feet.

With the diagnosis confirmed, Miriam is keen for treatment. She is seeking a resolution to her torment. Like most genetic conditions, however, there is unfortunately no cure. The cause of the symptoms is intrinsic to her DNA, part of her own make-up. And the bad news for her is that most cases of PE are very problematic to treat. Local anaesthetics like lidocaine act primarily on the sodium channel, and so it would make sense to try these drugs, and skin patches impregnated with local anaesthetic are often used to treat pain of nerve origin. But unfortunately for people with PE, the very same changes in structure and function in the sodium channel also cause changes in the site to which the local anaesthetic binds, rendering it ineffective in roughly half of all patients. And indeed, Miriam and her son appear to be in that unlucky half: the lidocaine patches do not work. So we try another drug, one more usually used for epilepsy and that also targets the sodium channels. And while this does not resolve her pain, it at least dulls it. The sensation of hot embers is replaced with a prickling heat, more tolerable, less overriding. 'I can sleep a little easier now,' she tells me when I see her next, 'and the attacks are less severe. It hasn't transformed my life, but it has made things a little better.' But the last time I see her, she is still wearing her sandals.

You might be forgiven for thinking that everything boils down to the sodium channel. While the mechanical principle – the opening and closing of molecular pores to allow electrical charge to move in and out of cells – is at the very heart of almost all biology, it does not explain everything.

You will remember Dawn, whose eyesight has been slowly taken away from her by benign tumours strangling her optic nerves; over the past couple of years, her arrival in the waiting room has become announced by her white stick tapping on the floor. But, despite the devastating blow of losing her vision, I am amazed by her cheerfulness, her breezy manner, whenever I see her, and by her husband's apparent stoicism in the face of family life becoming increasingly difficult. I can only imagine seeing your wife, fit and able only a few years before, gradually declining in ability and health. Whenever I ask how it is all going, I am greeted by a smile and a 'Not too bad!' as she folds her cane and takes a seat. Dawn has been an army wife since her teens. I suspect it is this resilience that has seen her and her husband Martin through some very dark times.

But today is different. As I call her in to the clinic room, the smile is gone, a grim look on her face. She looks ashen, her hair unkempt, and Martin follows, no longer with his usual relaxed demeanour. Without any of the usual small talk, and even before she has found the seat with her hand, she begins. 'I'm in agony! This is absolutely awful. I can't eat, can't drink, can't wash my hair. For the last few weeks, the pain in my face has been excruciating!' She continues, telling me of a pain – sometimes like a burning needle, sometimes electric in nature – in her left cheek and left upper jaw; a stabbing so intense she cannot tolerate it. If she is holding something, it drops to the floor with her shock at the severity of the pain. It is only momentary, a few seconds at a time, but recurs hundreds of times a day, making life impossible. 'The pain comes on whatever I'm doing, but certain things trigger it off – brushing my teeth, drinking or eating, sometimes even the wind or water on my face.' Hence the unkempt hair – she has been too afraid of worsening the pain by washing it. On one occasion, she was woken up by it, screaming with agony in the middle of the night. Martin initially thought she was having a nightmare. 'It is horrible, far worse than anything I have ever experienced. Like the most painful bit of child-birth.' But unlike labour, this has gone on for months, not hours. And as she comes to the end of her description, her last statement is worrying. 'I just can't go on like this!' She has barely ever complained about her vision, so this puts the intensity of her pain into context.

Pain and sensory symptoms do not result only from changes in the nerve endings, the receptors that detect sensation, or in the channels responsible for the electrical signal. Compression or irritation of the nerves themselves can give rise to a variety of sensations. The pain of sciatica, for example, will be familiar to many – a searing electric pain radiating from the lower back into the buttock, down the back of the leg and into the heel or foot. Worsened by walking or standing, it is excruciating and a common cause of disability. In sciatica, it is the nerve root – at the place where the nerves enter the spinal column to join the spinal cord – that is compressed. Between the vertebrae, the blocks of bone that form the spinal column, sit the intervertebral discs, cushions of soft tissue that allow the vertebrae to move relative to each other and that act as shock absorbers, preventing the vertebrae from grinding against each other. But occasionally, these intervertebral discs will protrude a little, having been pushed out of place, or may even rupture. And when this happens, the protruding disc may compress the nerve root, precipitating the pain of sciatica. Certain positions may exacerbate it and, if you are really unlucky, even coughing, sneezing or straining may cause a transient worsening of this bulging, resulting in severe exacerbations of pain.

But for Dawn, the cause of her pain is not a disc. A quick look at her MRI scan provides an explanation – for deep within the recesses of her brainstem sits another meningioma tumour, right next to the trigeminal nerve, the nerve that supplies facial sensation. The tumour has been there a while, not causing any problems. But as it has grown it has displaced and stretched the trigeminal nerve, triggering the severe facial pain she is experiencing, known as trigeminal neuralgia. The irritated nerve is firing spontaneously, without any injury or pathology in the face itself. And because of disruption of the nerve, there is 'cross-talk' within it, like the conversations in a crossed telephone call. Sensory stimuli like the wind on her face or her toothbrush gently touching her gums trigger the pain pathways, causing pain from sensations that should not be in any way unpleasant – even breathing. And the pain really is other-worldly – so overwhelming, so all-consuming, that I have on occasion heard people say, 'I can't think about anything else. There are times when suicide seems like a good option.' I have seen sufferers dehydrated and malnourished, such is the fear of triggering the pain by drinking or eating.

For many people, trigeminal neuralgia has no obvious underlying cause. It is sometimes caused by a blood vessel, lying slightly outside of its normal location, making contact with the trigeminal nerve. In severe cases, this blood vessel can be dealt with surgically. But the location and depth of Dawn's tumour, as with the one compressing her optic nerves, means that surgery is not an easy option. The knife would be potentially catastrophic and life-threatening. The neurosurgeon also looking after her agrees. We are left with the option of trying drugs to suppress these pain-generating impulses, and we start her on a cocktail of medications to alter the activity of the ion channels transmitting these signals. Over the next few months, while the dosages constantly increase, her trigeminal pain continues unabated, although occasionally she has brief windows of relief. I begin to dread her visits, feeling powerless to help, as the agonizing zaps continue. 'The vision, my other problems, are not a major issue at the moment. This pain – it colours everything,' she says. I send her to a specialist clinic, for alternative strategies, and she undergoes a number of procedures, with minimal results.

It becomes increasingly untenable to leave Dawn like this. Everything we have tried does little to ease her pain, now the dominant feature of her life. After lots of discussion, among the doctors involved and with Dawn, surgery is proposed: a removal of part of the tumour, an attempt to decompress the nerve. She understands the risks, and the possibility that this, too, might fail. Speaking of the surgery, afterwards, Dawn says, 'I didn't really have a choice. I had to try and relieve that pain. The decision was in desperation.'

After regaining consciousness, Dawn is confused for about three days. As her faculties return, she realises that she is pain-free. But she has not emerged from surgery unscathed. The left side of her face is numb, the searing agony replaced by nothingness, an inability to feel anything. And, when she talks, she slurs her words. When I see her shortly after the operation, she sounds half-cut, several gin and tonics down the hatch. She is clearly relieved to be free of the pain, but anxious about the complications of the surgery. With time, however, her speech improves, and nine months after surgery, listening to her, she sounds like she always did. The numbness has remained, however. 'Hopefully, the numbness will improve, but the horrific pain isn't there. It drags you down. It was definitely worth taking that step and having the surgery.' A price worth paying.

As I have shown you on many occasions, disruption of our senses can occur through damage or dysfunction at any point along the pathway from sensory organ to brain. The sense of touch is no exception, but neither the loss of touch, nor an excess or amplification of touch-sensation, are always due to such damage or dysfunction. Through our nerves, the streams of information from our skin continue on their journey through our spinal cords and to our brains, but, occasionally, certain surprising sensory symptoms can result from problems within the central nervous system.

As Abdul (not his real name) enters the room, his anxiety is etched on his face. The space around him vibrates with his heightened worry, though physically he looks fine. He is young, in his mid-twenties, dressed in a tracksuit and of

Middle Eastern heritage, and is trailed by his equally anxious mother. I invite them to sit down and ask Abdul what is wrong. Within thirty seconds my heart sinks, as he unleashes a torrent of words before he has even taken his seat. The act of coming through the door, being in front of a doctor, has caused the dam to crumble. I catch 'Google' – my heart sinks further as I contemplate every doctor's nemesis in this situation – as well as 'water' and 'legs', and tears well up as he continues to talk. After a couple of minutes, I have no option but to interrupt him and ask him to slow down and to start at the beginning. He pauses for breath, wipes the tears from his eyes and begins to tell me his story.

He first noticed a strange sensation in his left leg a few weeks ago, like water running down the inside of his calf. But over the past few days, this sensation has spread, involving his right foot and occasionally his thighs, too. And this feeling of liquid has intensified to such an extent that sometimes he pats himself down, convinced that he might have lost continence, urine trickling down his legs. We talk through his symptoms and I ask him about other issues, like bladder and bowel function, vision, the arms and face, all of which are normal. Then he continues, 'And I looked up my symptoms on Google, and it came up with multiple sclerosis!' The tears recommence, and I see his mother's eyes glisten in synchrony. When we finally get through his medical history, with multiple interruptions, I examine him to find all is normal. But despite this, I fear that, on this occasion, Dr Google may be correct.

As we have seen with Irene and her loss of taste, the inflammation seen in MS usually causes a loss of function – in the form of numbness, weakness, loss of vision, and so on. But the loss of myelin, the insulating protein that wraps itself around nerve fibres, as happens in inflammation, may result not just in a loss of function; it can also produce 'positive', function-gaining symptoms. Imagine a tangle of electrical wires, all isolated from one another by their rubber insulating sheaths. If the wires' outer coatings become frayed or stripped away, the wires short-circuit, sparking furiously as electrical current jumps from one wire to another. An impulse that should lead from one place to another instead ends up in an unintended destination, resulting in confusion. And this is precisely what can happen in MS. Sometimes these impulses are triggered by a mechanical stretch of the nervous system. Patients with inflammation in the spinal cord in the neck will sometimes complain of tingling down their back and to their legs when they bend their neck forward, triggering an already irritated area – a symptom called Lhermitte's phenomenon. But other additional sensory phenomena can also arise: a gentle pinprick to the thigh resulting in tingling throughout the leg; the stroking of the back of the hand with a cotton-wool ball producing a strange sensation in the cheek, as if the delicate filaments of a spider's web have settled on the face. This cross-talk of highly organised circuits and pathways confusing the nervous system generates sensations from nothing, or causes misinterpretation, essentially illusions or hallucinations of touch.

Abdul's symptoms are certainly very consistent with MS. He describes a progression of symptoms over a number of weeks. He is young, and although his genetic origin is the Middle East, where MS is less common, he was born and bred in London; a curious feature of this condition is that the risk of developing MS is connected with geographic latitude in younger life – the further from the equator you are brought up, the higher your risk of having it. There are many theories as to why this might be the case – genetic variation, levels of vitamin D, possible infectious agents – but none are proven. In any case, Abdul's symptoms, affecting both legs, would certainly be consistent with a problem in the spinal cord.

With more than a touch of guilt, I play down Abdul's concerns for fear of escalating another onslaught of anxiety. I know that if I explore this now, our consultation will come to an abrupt end, at least from the perspective of achieving anything useful. I tell him that it seems likely he has a patch of inflammation in his spinal cord, but the cause of it is unclear. Occasionally inflammation may be triggered by an autoimmune attack in response to a viral infection, in which case it is unlikely to recur. I feel duty-bound to confirm that MS is in the mix as an explanation, but stress to him that the common view of MS being a life-threatening or devastating condition needs to be corrected.

Within a few days, he has an MRI scan. On the images I can see a large abnormality within the bottom end of his thoracic cord – the part of the spinal cord at the chest level – and further areas of damage in his brain. It all looks very much like MS, but what the scans do not show is that these patches of inflammation are of different ages. But before I can even tell Abdul the results, before the scan has even been reviewed by my neuroradiology colleague, my secretary receives a phone call from him, in tears, his terror at the possible diagnosis overwhelming him, and my efforts at trying to reassure him are a miserable failure. Once I have the results to hand I call him back to talk him through them, and once again I tell him that, while it could be MS, we need to see if the inflammation is recurrent or a one-off. I organise a repeat scan in three months' time, to see if there are any changes. But within a few days, Abdul is back on the phone, asking about a lumbar puncture. Analysis of the cerebrospinal fluid, the liquor bathing the brain and spinal cord – removed through the insertion of a long needle into the lower back – can help confirm the diagnosis of MS, but these days it is more typically done when the diagnosis is in doubt.

The inflammation settles down and his symptoms resolve. He is left with some minor tingling in his thighs, an echo of the inflammation in his spinal cord that has damaged the tracts of nerve fibres from his legs to his brain. His fate is going to be decided in the MS clinic, where a decision will be made about whether to start him on medication to prevent further relapses.

The reality is that all of this – the sodium channels, the nerve fibres, the columns of fibres ascending in the spinal cord – forms a prelude to the main event. As we have seen throughout the preceding chapters, the perception of our senses – the awareness of sensory inputs – happens in the brain, not in the nerves or the spinal cord. Many organisms, even single-celled creatures like amoebae, can sense the outside environment, moving away from noxious stimuli or towards food sources. So, in a way, they feel too; they respond to sensation, to the sampling of the chemical or mechanical world around them. But this reflex-response to the environment is a world away from conscious sensation. When *we* feel touch sensation, it is so much more than a reflex. We ascribe meaning to these sensations, interpreting what our skin tells us in the context of the internal world and the wider, external one. When we hold a 50p piece in our hand, we recognise it as a coin, and understand its utility. When we pet a dog, the sensation of its soft fur and warmth is not only consistent with what we expect a dog to be, but is also associated with an emotional quality, a sense of comfort and contentment.

And when we catch our thumb with a hammer, this does not simply result in a rapid withdrawal of our hand; the pain has an emotional aspect to it too. As we have seen, pain itself is not represented in one single brain area. There are multiple areas, each involved in subtly different aspects of the experience of pain – the location of it, the emotional component to it, even the physical changes such as increased heart rate, blood pressure and breathing rate, preparing us to run away or take other forms of action in the context of pain.

Our perception of our senses is crucially influenced by our attention. The brain is unable to process and make sense of absolutely everything in the world around us, so multiple processes act as a searchlight in the darkness, illuminating a small area of our sensory world, intensifying the detail. We are all familiar with not hearing something in our auditory environment until our attention is drawn to it, when all of a sudden it becomes rather obvious; or with the sound of a dripping tap, suddenly impossible to ignore once you have noticed it.

But attention is also an important factor in the perception of touch sensation. Sit still and quietly, and think about your feet. Pay attention to the feel of the socks on your toes, the pressure of the shoe-leather across the top of your feet, the feel of the floor on the soles. Sensations previously imperceptible come to the fore, like the blurry background of a visual scene suddenly brought into focus.

And this shifting of the attention spotlight is also at play when it comes to pain. Focusing on a painful stimulus exacerbates the perception of pain intensity, but being distracted by other mental tasks decreases activity in the areas of the brain responsible for localising pain. Treatments like mindfulness-based therapy aim to either distract you from your pain, or to help you understand the pain in the absence of its emotional component.

And as with all our other sensations, we are constantly analysing what we are feeling and comparing them to our expectations or previous experiences of the outside world. This concept of our internal model of the world, of drawing conclusions from our senses, is based on our predictions of the world around us. For all of us, there is sometimes a disconnect between reality and perception of sensation. Illusions are the most obvious example, and they exist in the touch sphere, not only in the visual or auditory. One famous phenomenon is that of the 'cutaneous rabbit'. Ask someone else to close their eyes; now tap them six times on their fore-arm – three times at the wrist, followed by a pause of two seconds, then three times at the elbow. They will likely feel successive taps moving up their arm, like a hopping rabbit – a potent example of touch information being interpreted in the context of expectation.

As if she has not had enough to deal with, Dawn has yet more symptoms as a result of her meningioma tumours. In fact, looking back at her medical records, I am reminded that the reason that I first met her, over a decade ago, was for some-thing entirely different that she has suffered from since long before the trigeminal neuralgia started. Several times a week, sometimes several times a day, Dawn experiences something unusual. 'It begins with a sudden wave of nausea. Then I get some tingling, always in the left shoulder,' she says. 'Within a second or two, the pins and needles spread to the left arm, the leg, the face, in fact the whole of the left side of the body. It then stays with me for between ten and thirty seconds before subsiding.' The tingling in itself is not painful, but, she continues, 'It is really distracting. I feel I just have to stop talking. Afterwards, I feel a little weak and off-balance.' On discussing these episodes with her, it is unclear whether she simply cannot talk when they occur or if she is so disturbed by the sensations spreading through her left side that she cannot maintain focus. Before the onset of her trigeminal neuralgia, it is these episodes and her visual loss that she found the most distressing. The visual impairment is a persistent problem, but these episodes, arising from nowhere, untriggered and unexpected, always throw her off-kilter, and she lives in the expectation that they could happen at any moment.

The identical nature of these episodes, and the rate at which they unfold, are absolutely typical of epileptic seizures – abnormal electrical activity spreading across the surface of the brain. Like Susan's visual seizures, in Chapter 2, they are not generalised convulsions, full-blown seizures causing unconsciousness, shaking and a risk of biting the tongue. Just listening to Dawn's description tells us where in the brain her seizures are occurring. The tingling in the left shoulder, followed by a rapid spread to the arm, the face, then down to the whole of the left side of the body, implies that her seizures are affecting the right side of the brain, in the primary sensory cortex. This area is one of those regions that identifies the anatomical location of the origin of a painful stimulus. But it is not just pain that is encoded here; it is the other sensory modalities too. It is here that sensation, at its most basic level, is processed.

This small strip of cortex runs from deep in the midline of the brain and wraps round to the outside, in the region of the ear. Imagine a diagram of your own body traced out along the length of the motor strip. At one end, extending into the sagittal sulcus – the midline dividing chasm between the brain's right and left hemispheres – are the foot and the leg, and, as we climb out of the sagittal sulcus, the hip, trunk, and neck. Moving further round and out, are the shoulder, arm and hand. But the sensory homunculus – meaning 'little man', referring to the representation of our own body in the sensory cortex – is not true to scale, nor is it entirely true to form. Those areas of our body that are more sensitive, with a higher density of receptors and nerve endings, where more detailed perceptions of sensation are more important, are allotted larger areas of cortex. This means that, in a homunculus diagram, areas like the hand, face, lips and mouth are hugely distorted, massively out of proportion to less sensitive areas like the trunk or thigh. In fact, the latter are represented further along the sensory cortex, beyond the hand area, reaching towards the temporal lobe.

If you think about your own experience of sensation, this distortion of how our own bodies are represented makes perfect sense. Perhaps you have been bitten by a mosquito on your back. You can feel the itch, and you know that somewhere near your left shoulder blade one of these cursed creatures has left its mark, which is now irritating you as you lie in bed attempting to sleep. But if you try to reach round to find the bite, it takes some doing, beyond the simple contortion of your arm; it may in fact take quite a while to find the exact location of the bite. But if you have a splinter in your finger, you will immediately know precisely which area of the finger it is in. The resolution of sensory detail varies hugely between your finger and back, due to both the density of sensory organs in each area and the representation in the sensory homunculus.

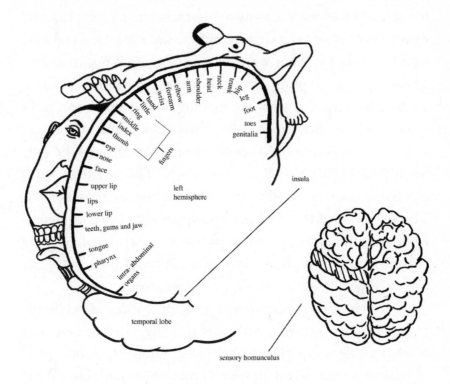

Figure 6. The sensory homunculus. Bodily sensations are processed by the primary (somato-)sensory cortex, but the body is represented in a distorted fashion, with more sensitive body parts such as the face and hands represented by larger areas of cerebral cortex.

From Dawn's description, it seems that her seizures are starting in the shoulder area of the sensory cortex, rapidly spreading, in one direction, down into the arm area, and in the other direction, to the trunk and leg areas. I go to her most recent MRI scan to look for a meningioma sitting over the right sensory cortex, irritating the underlying brain tissue. But despite the multitude of tumours gradually displacing and disrupting the underlying tissues, there is none in this location. As I pause for thought, it occurs to me that she has actually told me the answer – because the tingling that she experiences is not the first symptom she reports. 'The first sign I have is a sudden wave of nausea that comes over me. Then I get the tingling in the shoulder.' When I look again at the scan, there is another meningioma deep in the right temporal lobe. Seizures arising here often result in nausea, the electrical activity disrupting the nerve supply to the gut. So, it seems likely that this is the origin of her seizures, with electrical activity rapidly spreading to the sensory cortex. The activation of these areas by disorganised electrical signals results in the strange and unpleasant sensation of tingling passing over her body, a clear illustration of how sensation can exist purely in the mind, or, rather, the brain.

Over the years, a combination of drugs has largely controlled Dawn's seizures, but she is still left with the fading of her last vestiges of sight and her facial numbness. During the regular appointments with which I touch base with Dawn, my overwhelming sense is one of failure – the failure of modern medicine to deal with this most complex of problems. The cattle-prod-like jolts of pain torturing her day and night, and these sensory seizures, may have finally yielded to the endless prescriptions of medications and surgery, but her vision continues to worsen. And yet, despite this, for the most part Dawn is still smiling, refusing to relent to her condition.

The individuals in this chapter show that an illusion of touch may have its origin at any point in the sensory pathway – at the receptors on nerve endings, the nerve fibres themselves, the circuitry within the central nervous system conveying these messages to the brain, or the sensory cortex itself. This concept is common to all the senses; impairment or confusion of our senses may not be simply a result of damage to our sensory organs – our eyes, our ears, and so on – but may arise from any point in the entire machinery of our senses.

As with vision and the 'what' and 'where' pathways, or the link between olfactory centres and regions of the brain involved in memory and emotion, the sensation of touch needs to be put into context to have meaning. It needs to be integrated with our emotions, our sense of our own bodies, our other senses. We may feel a smooth, cold object in the palm of our hand, a flat disk at one end, and straight protrusion at the other, but we also know it represents a device for unlocking a door: a key. To sense a trickle of cold down our cheek is meaningless without understanding that it has begun to rain. And the persistent dull pain of toothache has an emotional quality to it, a sense deep inside us of unpleasantness. To ascribe meaning to all these experiences requires these sensory experiences to penetrate through to more widespread parts of the brain, those that process memory, emotion and the sense of our own body in space.

And no region of the brain is more illustrative of this perceptual experience than the parietal lobe, immediately adjacent to the sensory cortex. Among its many functions, the parietal lobe can be considered the seat of consciousness when it comes to sensation, where higher-order analysis confers meaning to all these sensory inputs. Damage to the parietal lobe results in a variety of problems: an inability to know where our own bodies are in relation to the world around us, difficulty recognising objects by touch, or even a lack of attention to ourselves or the world outside. Patients with parietal lobe strokes may exhibit inattention to touch or vision on one side, eating only one half of the plate of food in front of them or noticing touch only to one arm and not the other. And it is not just perception of the space around or within us that can be affected, but the memory of that perception. One of the most famous examples of this was a Venetian artist who suffered a stroke that affected his parietal lobe. He was asked to imagine himself standing on the south side of St Mark's Square and to paint what he could see in his mind's eye. He drew a remarkably accurate representation of the view, but only painted what was on the right side of his 'vision', leaving the left side blank. When asked to imagine himself to be on the north side of the square, he once again drew half the visual scene, this time the other side of the square. The visual memories were present and correct, but somehow he could not access the scene as a whole.

Perhaps there is no more striking phenomenon to exemplify the relevance of the parietal lobe to our sense of self than that of the phantom limb. The experience of abnormal sensations or pain emanating from an amputated limb has long been recognised. The US Civil War surgeon Silas Weir Marshall describes in a short story a soldier anaesthetised with chloroform for the amputation of both legs, which had been horribly mangled and rendered unsalvageable through battle. In the words of the soldier, Weir writes:

> [I] was suddenly aware of a sharp cramp in my left leg. I tried to get at it to rub it with my single arm, but, finding myself too weak, hailed an attendant.
>
> 'Just rub my left calf,' said I, 'if you please.'
> 'Calf?' said he. 'You ain't none, pardner. It's took off.'
> 'I know better,' said I. 'I have pain in both legs.'
> 'Wall, I never!' said he. 'You ain't got nary leg.'
> As I did not believe him, he threw off the covers, and, to my horror, showed me that I had suffered amputation of both thighs, very high up.
> 'That will do,' said I, faintly.

The phantom-limb hallucination – feeling that the missing limb is still there, weeks, sometimes decades after the loss of a body part – seems to affect almost everyone with an amputation to some degree. The symptoms that people experience vary greatly. For some it takes the form of pain, perhaps as cramp associated with an uncomfortable position they might have assumed before the amputation, or the feeling of their fingernails digging into the palm of the hand. For others, these sensations can be non-painful, like itching, heat or cold. And in a few, there remains a sensation of movement, either voluntary or involuntary. People have reported feeling able to wiggle individual 'fingers', to reach out to pick up the telephone or even to move the 'arm' in an effort to ward off blows or break a fall. At its extreme, the phantom-limb phenomenon can have some unfortunate consequences. One patient who underwent a leg amputation reported, 'I got out of bed and fell one time. Even when I had the [false] leg on before and would take it off, I still couldn't remember the amputation for two weeks after I was out of the hospital. And, I would try to walk and couldn't. I would just fall. I couldn't fathom I didn't have a leg there.'

The phantom-limb phenomenon is a fantastic illustration of the fact that the parietal lobe is home to a representation of our bodies, a body map. Very similar to the sensory homunculus, but a map of the concept of our bodies where they are in relation to the space around them and to each other body part, rather than specific touch sensations. And the various manifestations suggest that this body map receives inputs from our sensory systems, in the form of pain, touch and temperature, but also the sensation of proprioception – interpreted from signals from our joints, ligaments, tendons and muscles – indicating the position of our limbs. But there are further inputs, such as vision. Consider an experiment called the 'rubber hand illusion'. Imagine you are seated at a desk, your left hand hidden from view under the table. On the table top sits a rubber hand placed roughly above your hidden real hand. If the rubber hand is stroked with a feather, you are likely to actually feel the feather on your own hand, and you will feel that your hand is actually in the exact position of the rubber hand. This literally mind-bending illusion illustrates the complex streams of information that are utilised to define our perception of our own body.

But this body map is vulnerable not only to the loss of an arm or a leg. I have been seeing Laura (not her real name) in my epilepsy clinic for a number of years now. A young woman in her mid-twenties, over the past few years she has been plagued by occasional full-blown convulsions, one every few months. Even-tempered and low-key, she always downplays her seizures, and is very matter-of-fact when she describes her life being punctuated by these dramatic events. For someone who has never had a seizure, it can be hard to comprehend trying to live your normal life – to work, socialise, have a family – in the knowledge that any moment, without warning, you may lose control. At the flick of a switch, you lose consciousness, waking confused, perhaps having been incontinent, your colleagues or family members witnesses to a violent and awful event. We spend our waking lives trying to establish a degree of power over what we eat, what we wear, how we appear before others, and epileptic seizures instantly destroy this autonomy.

For the past year or two, we have been trying to establish control over Laura's epilepsy, all the while mindful of the fact that Laura wants to start a family. We need to get her off the most hazardous drugs during pregnancy. After several months of adding in new medications and weaning her off others, I see her again in my clinic. She walks in with a broad smile. 'I have not had any convulsions for several months,' she tells me. I breathe an internal sigh of relief. It has taken us a while to get to this point. 'But I am still having seizures!' she continues. I ask her what she means. 'Well, I am no longer losing consciousness. Those convulsions where I wake up on the floor, confused and sore – they have stopped. But I am still having the prelude to these events; the warning that I had still happens, but without the full-blown convulsions.' In my delight at her announcing herself to be free of her convulsions, I have forgotten the details of her episodes, and she reminds me. 'So, what now happens is that, at any time – at work, at home or out and about – I suddenly feel that my tongue swells. It feels far too big for my mouth, as if I have a cannonball in my mouth. Then, within a few seconds, my head begins to swell too. I feel like it has grown to about four times its size. I can literally feel it expanding, like a balloon on the end of a pump. No matter how many times it has happened, how used to it I am, the feeling of my brain, my eyes, my nose, growing bigger, it is so odd.' She has had these feelings before, usually as the herald to her convulsions, and so these hallucinations of her body shape are tinged with fear of an impending fit. And because, in the past, they have been followed by a major convulsion, her memory of this feeling is hazy. But over the past few months she has become habituated to this distortion of her body shape in isolation, and the anxiety associated with them has lessened. She has noted these feelings in more detail. 'It is very hard to describe, not something that many other people have experienced. Imagine that your body has no fixed shape, that it can stretch or grow. I view myself to be a little like a Marvel character, bending and stretching to wild proportions. In my mind, I know that this is not real, but it feels so real, as if my head is filling the room around me. And I tell myself that this cannot be true, this is not really happening, but it is very difficult to argue with your senses.'

And the cause of Laura's experiences? Her seizures originate in the parietal lobe, the seat of our own body image, the source of our representation of ourselves. While the new medications have stopped the abnormal electrical activity that underlies seizures from spreading to the whole of the brain, they have not succeeded in dampening down the small sparks in the depths of her neurological body map. Like firefighters dampening the vegetation around a burning fire, a forest inferno has been prevented but glowing embers remain, rekindling a flame every so often. Over time, we finally put out the flames completely, and she becomes seizure-free, able to lead a normal life.

Laura's case is a powerful example of how our sense of touch or sensation is not a pure representation of our external, or indeed internal, world. Our senses are an amalgam of our environment, of sensory triggers, and our internal processing. The outside world is meaningless without ascribing values to those inputs, without translation into our higher consciousness. Consider a balmy summer's evening in the garden. You sit with your eyes closed, the evening sun warming your face, the gentle breeze ruffling the hair on your head, the cooling blades of grass tickling the soles of your feet. You hold a chilled glass of wine in your hand; a ladybird lands on your leg. All incur sensory inputs that, when broken down to their most granular level, simply arise from a rich array of receptors detecting heat, cold, pressure, hair movement and so on. But your brain is a remarkable thing. It translates these inputs to allow you a fundamental understanding of what these electrical impulses actually mean, conferring a sense of reality to these basic signals. And even then, the story is not complete. As I imagine myself in that garden, glass of wine in hand, I can feel an overriding sense of pleasure, of relaxation, as these sensations inform my emotions, my memories, my experiences. It reminds me of childhood summers, of being carefree, without worries.

It is a virtual reality, a reconstruction of our senses, that we feel in the context of what has gone before. Laura and the others here vividly demonstrate this. Their conditions give us a clearer understanding of how the human machine detects, processes and determines the meaning of these sensory experiences; of the complex and highly organised system that takes raw data and, at every level, interprets the binary signals of these sensory receptors in our skin, becoming richer, more nuanced, more complex as this information percolates into our higher levels of consciousness, integrating our past – our previous experience and emotion of these sensations – and our future – our expectations – with our present. It is when the system goes wrong, when we experience alterations in our sensation, that we have the stark realisation that our perception of the physical world is a mere construct of our neurological systems. Subtle changes in neurological function, precipitated by disease, damage or genetic variation, fundamentally alter our world as we see it – or, in these cases, feel it. A chemical substance originating from algae, turning our sense of heat and cold upside down; a minor area of inflammation in the spinal cord resulting in feelings of water trickling down the legs; small disturbances of electrical function in the brain causing huge distortions in our sense of the world or ourselves: all amount to a recognition that sensation is a creation of the entire nervous system rather than simply the act of feeling the cold hard world around us.

9

THE PAIN OF SHEER HAPPINESS

'It was a pink sort of smell – a smell that seemed to
get bigger as you smelled it and then burst, just like
the popping of a bubble.'

Alexander McCall Smith, *Explosive Adventures*

By now, I hope to have truly convinced you that the way we experience our world is reliant upon the chemical and physical properties of our bodies, and more specifically our nervous systems, as much as on the physical properties of the objects around us. Our senses, the conduits of the exterior to the interior, are not set in stone. The illusions of everyday life – its sights, sounds, feel, and so on – are small illustrations of the methods or shortcuts that we take to construct our reality. And when things go wrong, be they tiny chemical changes (through poisoned fish, for example), acquired injuries or genetic changes from birth, these can fundamentally alter our understanding of our external world. While some of the people you have met in these pages have been affected by major disease, in others the dysfunction may come down to something as miniscule a change as a single letter in the genetic code that constitutes DNA – a simple typo that nonetheless defines the absence of a sensation, altering the core of the human experience. Each of our senses, the classic five senses but also those less well defined, like proprioception or movement perception, are highly vulnerable.

However, this relationship between the structure and function of our nervous systems and our comprehension of our environment runs much deeper than our senses. It is about much more than simply altering how we smell a rose or how we hear a voice. This relationship influences our internal world too. It influences the very nature of our understanding of what reality is. While this may sound rather far-fetched, there are people all around us whose experience of reality is very different from the rest of ours. This is not necessarily due to damage or disease, nor to some other sort of abnormality – they are normal individuals, with normal brains, normal genes. People indistinguishable from everyone else. Most have probably gone through life unaware of the differences between their 'reality' and the 'reality' of the majority – depending on what 'reality' really is. These people may only comprehend that they are different from others when directly asked about their experiences. Very rarely, we may even come across individuals who cross that 'reality' divide due to neurological insult or injury, undergoing a sudden transition in how they experience the world, joining those countless others out there who have been that way since birth.

When I meet Sheri, it is down a telephone line. Her voice is hypnotising. It is mellifluous; a gentle, comforting Canadian accent, the pace of speech easy, like treacle flowing off a spoon. Sheri is a painter in her late forties, living and working in glorious Vancouver Island, specialising in landscapes. She is very successful and has exhibited all over the world. When I later search for her on the internet, I am able to put a face to the voice. In her Facebook profile photo, she is perched on a stool, a dog at her feet, in front of what I presume to be one of her paintings. Her face is framed by chin-length light hair, a black hat with a thin brim adorning her head, her eyes gazing directly at the camera. Her paintings are stunning – large canvases of the Canadian wilderness, huge skies, occasionally more abstract accounts of the sky at night. But when we have our chat, I do not yet have a picture of her, or her art, in my head – only the voice, her beautiful voice.

At the age of twenty-nine, life for Sheri was pretty ordinary. She was studying for a masters in art. Her spare time was filled with water sports and she was taking some marine training to obtain boating qualifications. 'I was very busy, but it was just a regular, busy life,' she recalls. On the day that changed that life, she was practising lifeboat drills in a pool. She was standing by the side of the pool when she suddenly fell and crashed into the poolside wall. 'I couldn't feel the right side of my body. Luckily I was surrounded by first responders and I let the gentleman beside me know that I was having troubles. He lowered me to the ground and started going through a battery of tests. He did my vitals and noticed one of my pupils was larger than the other. I was conscious the whole time this was happening. It turns out I was having a stroke.' She talks about it dispassion-ately now, but it must have been terrifying.

Fortunately Sheri was in the city of Vancouver and the hos-pital was a stone's throw away, just up the hill from the pool. A few investigations confirmed the cause for her symptoms, and she was quickly admitted to intensive care and put on blood-thinning medications. She was confined to bed for a week, but could not have done much anyway, had she been allowed to get up and about. 'I had double vision, so I couldn't see properly. I couldn't walk. I was quite confused. I just spent most of the time meditating.' She was told that she had had a brainstem stroke, which explains the double vision and the difficulty walking, but not the confusion. Putting things together, I assume she had clots that also shot up to the rest of her brain, and that it was not just the brainstem that was involved.

Talking to her now, she sounds entirely normal, and I assume she has made a full physical recovery. When I ask her how long it took her to recover, to my surprise she tells me that the process is still ongoing, some twenty years later. She says that to get back to walking took her about three months. 'It was a progression,' she tells me. 'Initially, I pulled myself along the floor and learned how to crawl again. Then walking, but you could still tell that I didn't walk normally. My mom helped me take my first steps – like a child all over again. I guess it took maybe eight months to get to the point where no one could notice any difference.'

I imagine it must have been a traumatic time. 'It was very challenging,' Sheri says, with a hefty dose of understatement. 'I was more worried about my mom, to be honest. For me, at the time, it was a little bit bewildering. I didn't understand what had happened. I couldn't conceive of what rehabilitation would be like. I didn't know if I'd ever walk again. I just didn't know what my life was going to be like. I just looked at it as though anything might be possible: "I'll just try everything and see how it goes."'

In the following months, as the dust settled, her physical recovery became more complete. But there was one particular issue that was disturbing her enormously. 'I went to the eye doctor a handful of times, saying, "There's something wrong with my vision. I can't see properly," and he'd test my vision. It was fine. I had twenty-twenty, perfect vision. And I just kept saying, "There's something wrong, I don't understand this."'

I ask Sheri to explain what it was she was experiencing. She pauses for a while, formulating a response. 'When I was at university, my visual memory was very good. For tests, I would memorise pages of textbooks or my notes. But I have no visual memory now. My mind is just blank.' She pauses again. 'I tried to explain it to my mom. If you go to a movie theatre, the film is projected on a screen. Say if, in your mind, there's a screen and you're able to project anything you want on that screen, your mind just naturally does it. We're not even aware of it. After the stroke, as soon as I would try and project anything on that screen, it would dissolve. It was like there was something wrong with the light bulb in the projector in my mind.'

Even now, I get the sense that Sheri struggles to really put her situation into words. She likens it to a form of internal blindness, a term that makes the most sense to her. She remembers being simply unable to explain it any better. 'So, I just gave up thinking about it or trying to explain it any more. I focused more on, "How do I live like this?" Because I used that tool, that ability, so frequently. It was tied up in my identity.' That capacity to conjure mental images was crucial to her ability to paint, to think in a visual format. 'For me, it wasn't just a tool that I used to remember things, or construct my world. It was who I was. I just felt like I wasn't even human any more. I couldn't imagine anything. And my imagination was where I lived.'

For years, Sheri did not fully understand what had happened to her. She had made an almost full physical recovery from her stroke, but the lasting legacy of it was in some ways its worst. The stroke had stolen her vision – not her ability to see the outside world, which was untarnished, but her visual imagination. Her facility to conjure visual images from memory had been completely wiped out. Her internal projection screen had gone blank. 'For me, going from having such a strong and vivid visual imagination to going completely internally blind was more traumatic than having to learn how to walk again or read. That was the most traumatic experience of having a stroke.'

About five years ago, Sheri got some answers. She was driving in her car, listening to the Canadian Broadcasting Corporation. 'There was a programme on about this condition that some people have, where they can't imagine anything in their mind's eye. I had this wave come over me; I had prickles go through my entire body and my jaw dropped, and I had to pull over. I was literally shaking, because I could not believe what I was hearing, and I thought, "This is what's going on. This is what I can't do any more." It was so profound.'

The voice that she heard on that fateful day was that of Adam Zeman, a neurology colleague of mine based in Exeter. Zeman had initially come across a patient, known as MX (not his actual initials), a 65-year-old retired surveyor who had rather abruptly lost the ability to visualise. Under normal circumstances, as he was dropping off to sleep, he would 'see' the faces of friends and family, events that had happened recently, or even buildings related to his work. But all of this had disappeared a few days after he underwent a cardiac procedure called an angioplasty. The procedure had involved passing a wire into a large artery in his groin and up to the coronary arteries, where stents had been introduced to expand narrowed blood vessels. MX recalled having some vague symptoms during the procedure – some tingling in the left arm and some internal head sensations – but scans and detailed testing did not reveal any other deficits in his memory or perceptive abilities, nor did they reveal any obvious cause. In addition to his loss of visual imagery, for a brief period he had even lost the visual component of his dreams, but this had returned and he was left without his mind's eye. Curiously, when tested by means of tasks requiring visual imagery, such as mentally rotating three-dimensional objects, he performed normally, if more slowly. It was his awareness of visual imagery that was impaired, akin to blindsight, the phenomenon of seeing without being aware, described in Chapter 5. Having conducted extensive analysis of brain function, Zeman was able to demonstrate clear differences between the areas of brain activation in MX and in normal individuals during tasks requiring visual imagery, confirming that MX's condition had an underlying neurological basis. Zeman and his team suggested that MX's loss of his mind's eye was related to a tiny stroke that must have occurred during his coronary angioplasty.

Zeman subsequently termed this loss of visual imagery 'aphantasia', Greek for 'without imagination'. However, it soon became apparent that this phenomenon had been recognised earlier. In fact, Francis Galton, a Victorian polymath and part-time inventor, psychologist and controversial eugenicist, had published a study in 1880 in which he had asked 100 adult men to describe the table at which they ate breakfast every morning. Of those hundred, twelve men were almost entirely unable to give any sort of description. They had always assumed that everyone else had the same inability that they had to conjure up visual images, that the term 'mental imagery' was metaphorical, not literal. In addition to providing the first description of aphantasia, Galton also demonstrated something else important: that aphantasia did not result only from a stroke, as in Sheri's or MX's cases, but could be found in normal individuals, and a high proportion of normal individuals at that. Modern-day studies of visual imagery have confirmed that, rather than aphantasia being a rare phenomenon, it is quite common, affecting up to 3 per cent of people. For most, visual imagery has been absent since birth, although typically people only become aware of its absence as teenagers or young adults, when conversations with others, or reading, makes them recognise the existence of the mind's eye, and its absence in their own lives.

For Sheri, the effect of losing this mental faculty has been devastating. I ask her how it has affected her on a day-to-day level. She begins to answer, and then briefly pauses. 'That is a very overwhelming question, actually . . .' She swallows, and continues. 'Before the stroke, I was a figurative painter. I painted figures in acrylic paint, which is very fast-drying. After the stroke, I couldn't paint in acrylics any more. I basically retaught myself how to paint in oils. The other thing was that I lost all of my visual memory of mixing paint. I'm having to relearn how to mix colours pretty much every time I go to paint again.' As a non-painter, it is hard for me to appreciate the impact of this, but Sheri's art is her life's passion, her purpose. 'It's really emotional to talk about this . . .' She stumbles, then continues. 'I don't really know how else to explain it. Every time I approach the canvas, it's like I'm not able to carry through what I was working on yesterday. I have to relearn a bunch of stuff every day.'

The process of painting has also changed substantially for Sheri. She does not paint people any more; she paints landscapes now and is more interested in atmosphere, trying to capture a feeling rather than a specific image. Since becoming unable to retain an image in her mind, she has taken photographs of landscapes to work from. 'I hold the photograph when I paint. The photograph is in one hand and my brush is in the other hand, and I'm constantly referencing it, but I can't hold that image in my mind.' Her stroke has changed both the actual paintings she produces and the experience of painting.

In the rest of her life, there have also been some consequences. She is unable to visualise objects that are out of sight; it is as if the existence of anything hidden from sight is negated. 'I don't have doors on my kitchen cupboards because I can't see what's inside. I have a door on my fridge, but I don't know what's inside the fridge. If I'm cooking, my fridge door stays open.'

Throughout our conversation, I am listening to her rich voice as well as her words. But there is an undertone of melancholy. I tell her that it sounds like she is still grieving, in mourning for her mind's eye. She laughs. 'Yeah, I have a little bit of trouble hiding that, don't I? It was a part of myself and my life that I really indulged in. You know, when you're having an off day and you can visualise something pleasant? Or, if your grandmother passes away and you can still visualize her and the memories that you had, and the times that you saw her? I don't have any of that. For me, it's quite a lonely place. I don't have that sort of visual imagery to keep me company in times of joy or sorrow. It's just always blank, and so it feels like I'm living in a bit of an abyss.'

It seems, then, that it is not only our classical, 'external' senses, the windows on our outside world, that are influenced by glitches in our nervous system. Our 'internal' senses – those mental faculties that allow us to internalise our world, to recall it and bring it into conscious awareness with eyes closed – can also be impaired. And there are further parallels with our classic 'external senses'. Most obvious is that, as with Oliver and Dawn – the former with his visual limitation present since birth, the latter acquired later in life – the impact of the loss of a sense depends on whether you experienced life with it or it has never been an intrinsic part of you. As Adam Zeman has described, many people have never had visual imagery; aphantasia has been present since the day they were born. Compare Sheri and her loss of her mind's eye with one of her friends' experience of the condition. Sheri says, 'I've actually discovered that one of my new friends lives with aphantasia. We met at a dog event – we have the same breed of dog – and I just felt like I could trust her right away. After we got talking a little bit, she said, 'Oh, I live with this too, but I've lived with it all my life.' The difference I notice between myself and her is how, for her, it's just the way it's always been. She's so calm and confident about it. It's part of her, whereas for me it feels like a disability. And for her, it doesn't at all.' Indeed, since having a name for her diagnosis, Sheri has met many others with aphantasia. There has even been an exhibition of artists all with aphantasia. Sheri says that, with some, she can detect evidence of aphantasia in their art; with others it is impossible to tell.

And as Francis Galton and Adam Zeman have shown, aphantasia is firmly on the spectrum of humanity. A small proportion of the population have no mind's eye, despite having normal brains. In fact, the power of the mind's eye itself is on a spectrum, with some individuals having an extraordinary ability to visualise; others less so. In our abilities to interpret the external, and indeed internal, world, we are all a little different.

Aphantasia therefore illustrates that our internal realities may differ from each other, not only as a result of disease but as part of the range of normal. But what about our external reality? Are there examples of people for whom their entire reality, rather than a single sense, is altered or different in any way? Of course, there are those individuals with psychosis, those with hallucinations and delusions that have a basis in psychiatric disease. As I touched upon earlier, one possible way of considering these types of experience relates to over-interpretation of sensory inputs, our own model of the world overpowering what our senses are telling us.

But there is another category of individuals, healthy and

'normal' in every way, whose reality is somewhat different from our own. James is one of these. He is a synaesthete. He experiences synaesthesia – the merging of two or more senses. For him, this involves a fusion of hearing and taste.

Now in his sixties, it has been part of his life for as long as he can remember. 'My very first memories of sound having associated tastes and textures were during daily trips to and from pre-school on the London Underground. I was aged around four at the time, and my mother was helping me with my reading and writing, so I'd spend the journey reading out and writing down the names of the stations and their accompanying tastes as we stopped or passed through them. After that, I moved on to the overhead carriage maps. I still have some of those battered notebooks from all those years ago and the name, food and flavour matches described are exactly the same as they are today. I can also specifically remember we used to have to recite the Lord's Prayer every morning in my school. And that used to bring on extremely strong bacon, textural tastes.' James recalls mentioning this to his mother, but she would dismiss it. It did not occur to him that not everyone had this experience of the world. When he began to mention it to his school friends, he was generally greeted with blank responses, disbelief or mild acceptance. But his synaesthesia never really caused him a problem – until he was about fifteen. 'We used to have to sit yearly exams, which we would have in big, echoey halls. We'd sit there and I'd be totally and utterly distracted by the sounds, because the windows were always open – it was always in summer; or by a pencil rolling across a desk and then falling off and clattering on the wood floor in the gym. It used to distract me so much. Even reading the questions was very distracting. So I asked my mum to take me to the doctor.' I can guess what a GP's response in the 1950s or '60s would have been, and James confirms it, laughing: 'He said that from an early age I'd always had a wild imagination, and this was just a phase that I would grow out of. That was his response.'

It was a harsh assessment by the GP, but understandable. Even now, some of the things James talks about are utterly incomprehensible, almost outlandish. Discussing his Tube journeys as a child, he says, 'My favourite Tube station was Tottenham Court Road, because there's so many lovely words in there. "Tottenham" produced the taste and texture of a sausage; "Court" was like an egg – a fried egg but not a runny fried egg: a lovely crispy fried egg. And "Road" was toast. So there you've got a pre-made breakfast. But further along the Central Line was one of the worst ones, that used to taste like an aerosol can – you know, the aftertaste you get from hairspray. That was Bond Street.' At another point in our conversation, we are discussing whether his synaesthesia led him to make different personal decisions. When he looks back, he notes that all his old friends had names that tasted nice. His attraction to women has also been influenced by their names. Meanwhile, one of his friends was married to a woman whose name tasted to James of lumpy vomit – he grimaces as he speaks her name. Needless to say, James did not understand his friend's choice of partner.

In fact, his entire family has their own specific flavours and textures. 'My mother is called Doreen. I call it a taste but it's more of an experience. It's like the brain freeze you get when you have very, very cold water or ice cream. And my father, whose name is Peter – he tastes like tinned processed peas. And my sister tastes of blackcurrant yoghurt. My grandmother tastes of very creamy, thick condensed milk. Her name is Mary. My grandfather, William: he had a taste of crushed-up aspirin.

That's pretty unpleasant.' I somewhat tentatively ask him what my name tastes of, a little fearful of the answer, hoping that I do not taste of lumpy vomit like the unfortunate wife of his friend. 'It produces a word-sound that gives me a taste and texture that's quite complex, but something similar to fudge. It's not very sweet; it's quite bitter. But having met you in person, and listening to your voice, puts another layer over the top of it and it sort of develops. The taste and texture of fudge is still there but there's an added sweetness, which is rather nice.' I breathe a sigh of relief, and tell him I can live with that.

I spend some time with James, trying to comprehend. It quickly becomes apparent that it is not as simple as hearing and taste; James is describing an actual mouth-feel, the texture, taste and odour of a particular substance. In essence, the flavour rather than simply the taste. And while this phenomenon occurs predominantly with word-sounds, this is not exclusively so. The example of the pencil rolling onto the floor in the exam room comes to mind, but James also finds that music evokes flavour in the same way. He says, 'Listening to music is fantastic because music produces mostly sweet kinds of tastes. Very textural tastes as well, and that's lovely. I couldn't imagine listening to music without having that as well, because it's all part of the experience.' Gentle piano music tastes of tinned pineapple chunks. Heavy metal is the chocolate on top of a digestive biscuit. 'I don't like live music so much. Jazz music, for example, is a bit of a nightmare, because the mix of tastes and textures in there is one after the other. It's like stuffing loads of food in your mouth.'

It is clear, however, that it is not simply a question of the sound of a word or music. There is a semantic element to all this. James has created a version of Harry Beck's original map of the London Underground with the names of the stations replaced with the flavours that James experiences with their sounds. Waterloo tastes of sparkling water, Kilburn tastes of rotten meat (presumably related to 'kill'), Holborn of burned matches ('born' sounding like 'burn', perhaps), Liverpool Street, not surprisingly, of liver and onions. There are several examples of these more obvious associations, but for the majority I cannot figure out the code.

And it is not even the actual sound that triggers flavours for James. He says that inner speech can have the same effect, and even looking at objects can precipitate this experience. As we talk, he is looking around his sitting room. His gaze settles on the television screen: 'I get the taste and texture of jelly whenever I see one,' he says (perhaps due to rhyming with 'telly'). There is a red armchair in the corner of the room. I ask him if it, too, is giving him a taste. 'Yes – it's a bit like the residue you get at the bottom of a jam jar. You know, that thick, gooey muck at the bottom, slightly liquidy.'

It all sounds extremely far-fetched, and I can understand James' GP's disbelief. But despite his vivid tales, there is evidence that James' experiences are very real, not just the product of a bountiful imagination. For a start, the flavours of words or objects are totally fixed. A look through his childhood notebooks confirms that the flavours of Tube stations now are the same as they ever were. The tastes of people's names, of Tube stations, are immutable. Of course, he may simply have a very good memory, but these associations are so numerous that James would be extremely hard-pressed to remember every single one. It is possibly persuasive, but not definitive of synaesthesia being a real phenomenon. For some synaesthetes, the rules governing their associations are so complex that they cannot voice them or even understand them; such rules have taken specialists in psycholinguistics years to unravel.

Even more convincing, however, are studies of the brains of synaesthetes. Using the almost ubiquitous tool of the neuroscientist, functional magnetic resonance imaging (fMRI), researchers have clearly demonstrated differences between synaesthetes and non-synaesthetes. For example, for those who report perceiving colours from words or letters, during language comprehension tasks the colour-selective regions of the brain light up. This definitively shows an underlying neurological basis for their experience of the world. Other studies have shown areas of hyperconnectivity within the brains of synaesthetes. There are clear expansions of connecting tracts between different parts of the cerebral cortex that presumably mediate this cross-talk between the senses.

A third line of evidence comes from genetic studies. These have identified areas of the human genome linked to synaesthesia, obviously implying a genetic basis to the phenomenon. This supports the observation that synaesthesia tends to run in families. Here, once again, Francis Galton pops up, having published a paper in *Nature* in 1880 entitled 'Visualised numerals'. While Galton does not use the term 'synaesthesia', this is clearly what is being referred to; he describes many cases of individuals who see numbers when they are spoken:

The writer is an office-bearer of one of our scientific societies: 'If words such as fifty-six be spoken, I most clearly, easily and instantly visualise the figures. I do so almost automatically. I perceive that when I speak the word "thousand" or hear it spoken, the figures at once group themselves together ... The figures are always printed; in type and size they resemble those commonly used for the headings of newspapers. I cannot, however, appreciate a back-ground, the figures appear simply in space.'

In the same paper, he goes on to describe individuals with family members who experience something similar, and specifically comments on the role of heredity in this trait; an early recognition of the role of genetics in synaesthesia.

As Galton hints at, synaesthesia does not purely involve the merging of sounds and flavours, but can affect other senses too. Valeria experiences a different form of synaesthesia, one that she has put to good use. Now in her mid-twenties, she is studying music psychology in London, while also working as a musician and singer. We meet in the Royal Opera House in Covent Garden, where she works part-time. She is bright-eyed, with a broad, ready smile, and speaks with the gentle sing-song intonation of a native Italian-speaker. I ask her when she first realised that she was different from other people. 'I was thirteen or fourteen years old. At school, we had to prepare a short film, maybe ten minutes, about a book. I was asked to do the music for it, because they knew I played piano. I remember talking to my dad about this, and I said, "I could make it in D major, because the cover of the book is green." He could not understand what I was talking about. I told him, "The cover of the book is green, D major is green. It would be nice to do it in the same colour." He was like, "We don't usually see colours when we hear music."' And that was Valeria's first indication that her world was different from that of those around her.

Her synaesthesia has always been there. 'You just don't think about it; you just have both at the same time. It's like when you eat something and you also smell it. It was the same for me with music and colours – it's like textures and sensations, when you talk to your piano teacher about how a piece of music feels, the things you say about the musicality and interpretation of a piece. But I never realised that it was just a metaphor, not actually real. When I realised it wasn't normal, I started to keep it to myself.'

As we talk, she begins to tinkle on the grand piano, the foyer of the Royal Opera House filling with music. She starts to play that original composition based upon the green book. As she moves through the piece, she describes how she experiences the music. She plays a chord, and tells me, 'This chord is very green and fresh. It almost feels like if you cut a lime in two and you touch the inside of it – that sensation, it's what I feel when I play this.' As she changes key, she says the colour has changed to blue. I ask her if she sees these colours only in her mind's eye, or if it intrudes into her real visual world. 'It's here,' she says, holding her hand up to the side of her head, 'in my peripheral vision. As if I had a light just behind my head, projecting this way. I don't see the light in front of me but I see the aura, all around my peripheral vision. If I try and look at it, it's difficult because it moves with me.' Another key shift, another visual experience: 'It's the blending of the two colours. So, it's as if the green was wiped away with a brush of blue. And this fifth,' she plays a final chord, 'it's a dot of yellow in the blue.'

And Valeria's experience of the music is not just auditory and visual. Music is a sensory experience for her, too; she feels textures on her skin. 'It can be either a good or a bad sensation. Most of the time I feel something kind of hugging me – the upper back, the shoulders and the arms. That's usually a good sensation. Whereas if I'm playing this,' – a loud, crashing, unpleasant chord – 'these chords are quite sharp and the colour is a yellow, but quite a stabbing yellow. And I feel it right here in my lower back. It feels slightly more uncomfortable.'

She stops playing her own composition, and moves to Debussy's First Arabesque. As she plays, she also paints a verbal picture, although for her of course it is a literal picture. She plays a single note, and tells me, 'You start with that dot of sound. It's a ball of sound and colour, and then it becomes more blue and much, much deeper. But it's not watercolours; it's very present. It's like oil on canvas. It's something much more real and intense, as if you took a brush, you put it in paint and you are adding one colour to the other. And they don't really mix. It's not like if you put the blue on top of the yellow you get green. You get layers of colours, as with layers of sound.' All this time she continues to play, and with each chord and each change in key she describes new colours and new sensations – bright oranges, indigos, purples, yellows, matched with feelings of warmth on the face, an ocean breeze, a stabbing sensation around her spine. I am reminded of son-et-lumière shows, where music is matched to light displays and fireworks. It must be amazing to have that, day in, day out, woven into the fabric of your life. Valeria seems to experience music on a totally different level. There are some pieces that for her do not quite work on a visual or sensory level, whereas some composers seem particularly apt at matching the sound of the music to its visual or sensory experience. 'I am convinced Debussy, Ravel, Alberniz – they were all synaesthetes. The fact that they put so many colours and so many variations in the music makes me wonder whether they were seeing reality with different colours as well.

'There is one sensation that is just utter beauty,' she tells me, when she has stopped playing. 'That's the sensation that I get when I cry with happiness.' I am expecting her to report some beautiful physical phenomenon, but to my surprise, she continues, 'And that's when I get the pain in my thumb. It's strange because it's a pain, but I feel so happy. It has happened very few times in my life, when I've been listening to very specific pieces of music. One of them is Beethoven's Ninth. There is one moment where my thumb is so painful, but that pain is the most beautiful thing. Everything is just so full of colours and textures. Actually, Ravel's Piano Concerto is another one. There is one moment when the oboe comes in. That's just perfection . . .'

While Valeria's hearing-induced vision is the most common type of synaesthesia, the condition comes in many different guises – visions generated by touch, tastes generated by words (as in James' case) or any other possible permutation of sensory crossover. It might, however, be a misrepresentation to consider synaesthesia as purely a merging of the senses. For some synaesthetes that is indeed the case, but for others it is not quite so straightforward. The triggers for these cross-modality experiences may not be raw sensory experiences, but higher-order cognitive constructs, with something more abstract generating these synaesthetic experiences. Consider James and his word–taste synaesthesia. For at least some of his pairings, it seems that the relationship is defined by the meaning or root of the words themselves. For him, there is a linguistic or semantic element to all this, not purely auditory. For others, certain words beginning with the same letter may elicit the same colour, despite the words being pronounced differently – e.g. popcorn, psychiatry, phone. Some synaesthetes experience different colours with different letters of the alphabet. For some it may be the sound of the letter; for others the geometry of the letters; for others still it may be that a letter, no matter how it is written – in whatever font, capitalised or lower-case – may precipitate the same colour, once again implying that the origins of their synaesthesia are more complex than the raw sensory experience.

Indeed, some of the experiences of synaesthesia seem to move significantly away from a simple merging of the senses. Some synaesthetes experience sequences like letters, numbers or months of the year as having different genders or personalities – the letter P representing a male or a sad person, for example. Others experience time mapped out in space, for example the days of the week laid out in an ellipse in front of them. Some experts argue that there are as many as 150 different types of synaesthesia, depending how carefully they are defined.

What is clear is that synaesthesia fundamentally alters the way people experience the world. And this is not a rare condition. Estimates of its frequency vary, but some studies suggest that as many as one in twenty individuals may experience elements of synaesthesia, although they may not always be aware of it. The vast majority of these individuals are 'normal' – normal brains, normal genes, no toxins, no pathology; they are normal human beings in every way. But their reality may be subtly, or in some cases not so subtly, different from others'.

If synaesthesia is so common and has an underlying genetic basis, then we need to ask why the genes that contribute to it have become so prevalent. What is the evolutionary advantage of having synaesthesia? For Valeria, one can understand how the possession of her synaesthesia might make her a better musician or composer, more creative, more gifted. She says, 'Being a singer and a musician, a big component of what I call my musicality is due to synaesthesia. Because naturally when you have preferences for a texture or a colour, you express that in your practice and in your playing, or your singing. And that definitely informs my way of being in the music profession.' Indeed, synaesthetes are more likely than the average person to be engaged in artistic pursuits.

However, it is difficult to immediately see how this confers an evolutionary advantage, a skill that is likely to improve the odds of survival for you and your offspring. But it may be that the creativity is a red herring, a side effect of having a hyperconnected brain. The presence of synaesthesia may make you better able to perceive or understand your external environment, and it is this that facilitates your survival. Imagine the sound of a twig snapping, signifying the presence of prey or predator. If that sound is accompanied by a flash of red in your vision, you can begin to see why that might put you at an advantage. Studies of synaesthetes have shown that when they hear words in languages they are unfamiliar with, they are better able to guess the meanings of the words just from their sound. So, if, as a prehistoric man, you are in a cave trying to develop a language system in your tribe, if you are a synaesthete you may be better at guessing what your compatriot's grunts mean. Better communication results in more successful hunting, gathering and security. This may translate in modern times to being extraordinarily adept with words. Similarly, other types of synaesthetes may have advantages in other sensory spheres.

The possible benefits of synaesthesia are likely balanced with disadvantages. It may enrich one's life, but it may also trouble you. It can be incredibly distracting. Valeria gives an example of this. 'So, let's say I have to learn a piece as a character, maybe in musical theatre. You have a very specific sensation about the song, but the character has a completely different one, and you have to be able to dissociate what is yours and what is the character's.' For her, if the colours or the feelings of the song are different in quality from what the character is trying to portray, this can be quite a challenge. James reports something very similar. As well as the experience of the exam hall in childhood, he describes many other examples. 'I can walk into a room and I'll get a taste,' he says. 'And if it is a particularly strong one, I find that very distracting. I'll have to look around to see what's producing that.' In a busy environment, the constant drip of mouth-feels and flavours, one after the other, like a never-ending buffet of

clashing foods, can make it very difficult to focus. When I talk to Julia Simner, one of the UK's experts on synaesthesia, she cites two other examples, both patients who are children. One young girl would struggle to breathe whenever people talked about hair, because she would feel it in her throat. Another had a pathological fear of the number four, because for her number four was a bully.

A more recent observation is that people with autism tend to display synaesthesia more frequently than usual. The nature of this association is not yet fully understood, but may have its basis in brain organisational changes or underlying alterations to the brain's processes of perception. But it is striking that a tendency to sensory overload is something that is frequently seen in individuals with autistic spectrum disorder. Many people with ASD find that too much sensory input, for example in a busy location with lots of sights and noises, can be totally overwhelming. It may be, in part, that synaesthesia relates to the excessive sensory experience of the world.

You may be asking yourself why I am telling you all this. Apart from the curiosity of these people's experiences, what is the relevance to previous chapters of this book? Well, throughout its pages we have seen the fragility of our relationship with the reality around us. Our senses, supposedly the conveyors of precise, accurate and sensitive information about our world, should perhaps be better considered as interpreters of data that may help us in the pursuit of our life's purpose: to find food, find a partner, procreate – to ensure the ongoing survival of our genes. What they tell us about the physical nature of our world may be limited, subject to abstraction or simply a form of shorthand, to facilitate our lives. And that this relationship is ultimately highly dependent upon the functional and structural integrity of our nervous systems – the mechanisms that capture this information from beyond our bodies and translate those

signals into something meaningful. As we have seen, disease states, injuries or other pathological processes can have a potent influence on our understanding of the world, but so can the range of normal human variation.

Consider this thought-experiment. Ten synaesthetes sit around a table, gazing at a red apple. All of these individuals are normal but will each have a different experience of that apple. One, for example, may experience the apple as having the taste of Coca-Cola as he gazes upon it. Another may feel a prickling on the back of his neck as he looks at it; the apple 'feels' prickly to him. A third may hear water flowing as she sits there, triggered by the redness of that apple; for her, the red apple burbles like a stream. And so on – each person's experience different, each person's apple real to them. Ten different realities.

I put this scenario to James and ask him which of the ten synaesthetes is experiencing the true reality. He replies, very matter-of-fact, 'They're all experiencing different forms of reality.' So, there are multiple realities, of which no single one is the only true reality.

Here's another thought-experiment. Imagine that every single human on this earth has Valeria's gift, that we all have her music–colour synaesthesia. Each and every one of us would see lime green with the D major key. All of us would see and feel the same things with the same pieces of music. The fact that sounds are colourful or full of sensation would be reality – the reality of the human experience. That reality would of course be a function of that hyperconnectivity of our brains, of the neurological changes that underlie synaesthesia. But that underlines the point that the lack of this link now – the reality that we currently accept – is also entirely dependent on our brains, in the absence of this hyperconnectivity. Our reality is a product of our bodies, a construct of our brains.

You may argue that my take on this is too simplistic and naive. You may counter that these scenarios of synaesthetes and multiple realities are flawed; that when one considers another world, one where almost everyone is red–green colour blind but a few can see in full colour, this does not reflect different realities, only different abilities to translate true physical properties into experiences. The perception of colour is different in these two groups, but in both cases reflect the immutable physical properties of reflection of light by the molecules of the objects we see. You may view this as intrinsically different from the synaesthesia scenario, where perceptions such as colours from music have no real basis in the physical world. However, if there is one message in this book, it is that our perceptions are often far removed from the physical. Our experiences and cold, hard reality can be almost entirely divorced, as with molecules of a particular structure and our experience of smell or flavour. Our experience of the world is a label for an interaction with our environment, a construct of our brains. So if we were all to be synaesthetes, and our synaesthesia were to follow rules – all right-angled objects smell of roses, all round objects smell of cheese – our perception of these objects would indeed relate to their physical properties.

Regardless of your take on this, for Valeria, James and others like them, there is something quite magical in the way they experience life when compared with those of us without this gift. As Valeria says, 'I am appreciative of my synaesthesia. I know it is something quite special, to be able to see colours when you hear music. It is like going to a museum or gallery and listening to the paintings. I know how special it is. Do I take it for granted? Sometimes, yes I do, because it has always been there.' She pauses briefly to think. 'There are so many ways of experiencing the world.'

EPILOGUE

THE TRUTH ABOUT THE TRUTH

'All our knowledge begins with the senses, proceeds
then to understanding, and ends with reason.'

Immanuel Kant

Failure is the human condition. Failure of body comes to us all
in the end. A few of us are lucky, but for most of us, failure of
body and mind is an accompaniment to life.

I have been fortunate thus far, and have escaped serious ill
health. But my body is no stranger to failure. I am extremely
short-sighted, blind as the proverbial bat, and have been wearing glasses or contact lenses since I was seven years old. At
school I played a lot of rugby, and until fairly late on, when
contact lenses became more mainstream, I spent hours every
week on the rugby field, flailing around in a blurry world.
Waiting to catch a ball kicked high into the air was largely
guesswork, the clear, oval form crystallising in my vision only
a split second before arriving in my arms (or landing on the
ground, as was often the case). My teammates were identifia-
ble only by the dark blue of their jerseys, and were otherwise
indistinguishable to me. The only saving grace was my size and
strength, which made up for my other deficits. But ultimately, I
knew that I would only have to put on my glasses and my vision

283

would return to normal; my experience of the world would be the same as my classmates.

I have endured other examples of failure of body too, some mundane and fleeting, others slightly more significant. There's the regular experience of sitting in a strange position, compressing my common peroneal nerve as it passes over the head of my fibula, the bony prominence just below the outer aspect of my knee. Realising, just a moment too late, that the foot I think I have placed on the floor as I have risen from my chair simply is not there. I have no knowledge of where my foot is, cannot even feel its presence, and I stumble and fall. Then there's the time I got a new bike, and had the set-up slightly wrong. Prolonged pressure on my hands and wrists from resting on the handlebars resulted in an ulnar palsy, the compression of one of the nerves supplying my hand. For several weeks afterwards, I was left with a deep, niggling ache in my arm, indeterminable in origin and not relieved by any contortion of my arm or hand. It was only when I went to turn the key in the front door or tried to chop an onion that I realised that the power in my right hand was missing. Self-examination also revealed some diminished sensation in the little finger, and I was toying with badgering one of my colleagues for some electrical studies when I saw that it was recovering on its own.

Despite knowing precisely what the problem was, and it only lasting a few short weeks, this episode was a deeply disconcerting experience – not because it highlighted my physical frailty, my vulnerability to medical issues just like everyone else, although that in itself was enough. Rather, it was the fact that it had taken me a couple of weeks to realise that the power and sensation in my hand was abnormal; that my own powers of perception could be so fragile as to not notice a deficit in my own body. I did not have the excuse of Oliver and his missing visual field, or Paul and his absence of pain, both conditions present since birth, both men having never known anything else. My problem had come on acutely, one moment having normal sensation and power, the next numbness and weakness. It led me to doubt my own faculties, my own senses, the veracity of my own body's witnessing of the world in question – a personal illustration of the gulf between reality and perception.

'Perception is nothing more than a controlled hallucination.' This is a commonly used sentence in the world of cognitive neuroscience. Essentially, our brains work as guessing machines, interpreting what is coming in through our senses in the context of our model of the world. What we perceive relates to our existing beliefs about the world, to how the information our senses provide us with interacts with our virtual-reality simulation of the universe.

The evidence for this is all around us, in the form of illusions, such as those I have described in this book. To pick up on just two, many of us are familiar with the experience of feeling our phone vibrating in our pocket when we're expecting a call. Despite repeatedly pulling the phone out to find there is no missed call, the illusion will persist. Then there's that viral internet meme, the black and blue dress – or was it really white and gold? The true perception of the colours of this dress may have been influenced by our expectation of lighting conditions – one study suggests that early risers are more likely to think that the dress is lit by natural light, while night owls tend to interpret it as being lit by artificial light. The colours of the dress in the photo were perceived accordingly, with morning larks more likely to say white and gold, and nightjars reporting black and blue.

These examples illustrate that our expectations of what is likely to be present or about to happen directly influence our perception of our world. And as we have previously discussed, there is a pressing need for this way of doing things. Without an element of prediction, the system would break down. In the absence of prediction, there are three insurmountable problems: the inherent delay of sensory information before it reaches our brains means that what we perceive has already happened; our nervous systems do not have the bandwidth to convey every single bit of sensory signal to our brains, nor do our brains have the power to process it; there is intrinsic ambiguity in any sensory information, and to resolve it needs a best guess.

And so, this idea of 'controlled hallucination' recognises the concept that while our senses are vital to understanding the world around us, our perception of the world is firmly rooted in our brain's own virtual reality. What the senses are telling us feeds into this simulated environment to help us understand these signals. When our senses clash with our own internal view of the world, this may give rise to illusions like those described above or, in extreme situations, when our inner simulation is more chaotic or frenzied, full-blown psychosis.

Even from this starting point, we already recognise that there is deviation between true reality – the cold, hard molecules around us – and our perception of it. What we perceive to be real is to some extent a figment of our own minds, a construct of the networks of neurones that constitute our brains.

But for some cognitive neuroscientists, this is too conservative an explanation, too literal a view of the world. There are some scientists who would go as far as to argue that we have absolutely no fundamental understanding of what reality actually is. These are not quacks, not scientific pseudo-babblers, but respected, eminent individuals. People like Donald D. Hoffman, a professor at University of California, Irvine.

Hoffman proposes that, contrary to our brains trying to represent reality for us, they have developed to actually *hide* it from us. Our minds construct a simplified or codified world to enable us to survive. Even as I write this sentence, it seems bonkers, totally ridiculous, no matter how many times I have read or heard his hypothesis, because it is so counterintuitive to what I 'know', what I experience on a daily basis.

The orthodox view, one that Hoffman does not share, is that, while we do not see reality as a whole, we do see those aspects of reality that we need to survive; so, when we look upon an object, that object is actually present, and there are aspects of it that we perceive – the red apple on the table does indeed have physical properties that confer redness or its shape. There are truths of the world that it is important for us to perceive, to enable our survival, and we perceive those fairly accurately. What we see is a reasonable summary of the atoms and their properties that constitute that apple. There are other truths about our world, however, that make little difference to our survival, and therefore evolution has not pressured us into seeing these truths. An excellent example of this is electromagnetic radiation. We perceive visible light, but this is only a very small range of the electromagnetic spectrum. We are surrounded by radio waves and cosmic rays, all of which are invisible to us. While they potentially cause us some harm, they are unlikely to kill us before we pass on our genes. There has never been any evolutionary imperative for organisms to perceive these rays.

But Hoffman's view is somewhat different. In fact, he describes this orthodox view as 'fundamentally wrong all the way down'. Hoffman actually questions the very nature of reality. He tells me, 'I think that our best scientific theories – namely, evolution by natural selection, quantum mechanics and Einstein's theory of space-time (general relativity) – are all pointing to the same conclusion. We've believed for centuries that space and time, or their combination into what we call space-time, is fundamental, objective reality. And that the contents of space-time, like atomic particles, are therefore part of objective reality. But our best science is now telling us that space-time is doomed, that space-time is not fundamental, and that we're going to have to look for some deeper understanding of reality that's outside of space and time.' Hoffman argues that although our brains create perceptions of space and time, of physical objects, these are not a picture of reality. He terms this a 'user interface', there explicitly to hide the nature of reality, to simply enable us to survive, while we are utterly ignorant of what we are actually doing in reality. He likens it to a computer desktop. When we click on an icon to open a document on our computer, what we see is simply a representation. The small white icon with the blue W of Microsoft Word tells us nothing about the internal wiring of the computer, nor the nature of the digital code, the series of os and 1s that constitute the encoded data. It is simply a representation that provides a useable format.

Hoffman's hypothesis sounds crazy, but is supported by some evidence. Nevertheless, I cannot get my head around it. I tell him it all sounds very redolent of *The Matrix*, referring to the trilogy of sci-fi films, in which Keanu Reeves plays the main character, Neo. Hoffman responds, 'Yes, but in *The Matrix*, when Neo steps out of the matrix, he steps into a space-time world. So, I'm saying something even more radical. If we could take off our space-time headset, we would be in the real world, which has no notion of space and time at all.' Our perceptions do not have the capabilities to show us the truth.

Hoffman's theory is the subject of a book, and perception

in general could generate another ten books. But I tell you all this to highlight that even the traditional, orthodox view of perception, which sits at the other end of the spectrum from Hoffman's, is that we are not necessarily aware of reality and that our brains tell us what we want or need to hear, see, feel, smell or taste. The truth is not necessarily what we perceive it to be.

Of course, the pathway to perception is a two-way street. Our ultimate experiences are a balance between sensory inputs coming in and our internal model of the world, a delicate equilibrium between 'bottom-up' and 'top-down' flows of information. But even our internal model, our virtual-reality environment, is fundamentally a function of our senses. Our internal model of our world is constantly tweaked, adjusted and fine-tuned, according to our experiences. This is how we learn to walk, talk, interact. It is even how we learn to see or hear. These are not skills we instantly possess from the moment we are born. Our senses are the windows to the outside world, the interface between ourselves and everything else, and that leads to our understanding of it. In the words of Aristotle, 'The senses are gateways to the intelligence. There is nothing in the intelligence which did not first pass through the senses.' The senses are the foundations for everything we are, everything we know, everything we believe, our values and our ethics. But, as illustrated by the extraordinary people whose stories I have been permitted to tell, these foundations are built on sand, not bedrock. A small earthquake, even a little tremor, in the form of damage or disease, can shake these foundations and make the walls of our reality tumble to the ground.

The stories of the individuals in this book also raise a philosophical issue. Firstly, they demonstrate some of the shortcomings in one of the core problems in philosophy of mind, the discipline concerned with what constitutes mental properties like perception and consciousness. The 'mind–body' problem, introduced some four centuries ago by René Descartes, tries to address the relationship between what constitutes our physical and mental being. For Descartes, the mind and body were made of different matter, each influencing the other but entirely separate. The nature of this relationship, between our spiritual mind – what comprises our free will, our mind, our soul – and our physical body remains an ongoing philosophical discussion, particularly when it comes to the nature of consciousness. I am, however, a physician and a scientist. I make no pretence at being a philosopher. In my view, it is likely that the physical nature of our bodies explains all aspects of our minds, that this dualism of body and mind is a false dichotomy. The individuals in this book, and their experiences of the world, potently illustrate that tiny physical changes can fundamentally alter not only our senses, but also our perceptions and our realities. They can influence our universe and influence our consciousness, directly demonstrating that our conscious mind is a function of our biology. I think specifically of Paul, whose entire world – whose understanding of the universe – is dependent on a tiny change in his genetic code. I empathise with him, I can rationalise his absence of pain, understand the impact, but can I truly comprehend what it is like to be him, to perceive the world in the way he does? I cannot for a moment put myself inside his mind – and all as a result of the change to a base pair in the sequence of his DNA.

As I write, we are in the midst of the second wave of Covid-19 in the UK. This morning I awoke to see a news story about one of the junior doctors at the hospital where I work being confronted by Covid deniers protesting outside as he left after a night shift. I am filled with anger at these morons, those who are anti-vaxxers and anti-scientists, the commentators who downplay the truth of what is happening – people so certain of themselves, of their world view, of their version of the truth, unquestioning, undoubting, unexamining of the evidence. This is, however, only a small part of the overall picture. We live in a cleaved world; increasingly polarised, with different world views. Brexit has fractured our society in the UK, albeit down predictable fault lines, and US politics is arguably worse, with its gross partisanship fuelled by the past few years. All around us, we see divergence of politics, of world views, of opinions.

There are, of course, ultimate truths (at least if you discount Hoffman's theory) – facts that do not rely on our perception or opinions, but are dependent on measurement or the physical rules of our world. For pretty much everything in the human sphere, however, it is perhaps not surprising that world views are discordant: 'I know this to be the truth, as I have heard it with my own ears, seen it with my own eyes.' Our understanding of the world is firmly rooted in our senses, but these senses are fallible and inconsistent, vulnerable to differences between individuals and to disease. While this is no excuse for idiocy, it is at least some sort of explanation for why each of us sees a different world and hears a different meaning. Certainty in our own beliefs, in our knowledge of the absolute truths of our world, is dangerous without a degree of questioning, or an examination of evidence to the contrary. Perhaps the faith we have in our own senses is misplaced. Our nervous systems are like the man behind the curtain in the *Wizard of Oz*, pulling the levers and pressing the buttons that make the magic happen. We ignore the basis of our reality at our peril.

GLOSSARY OF TERMS

affective Relating to mood or feelings.

amygdala An almond-shaped structure deep in the temporal lobe of the brain, part of the limbic system, with a primary role in emotional response, particularly fear.

aneurysm An outpouching of an artery, usually caused by weakness of the blood vessel wall. The resulting abnormality is liable to rupture, causing devastating bleeding.

anosmia The inability to smell.

antibody A blood protein produced by the immune system that acts to bind to and counteract alien substances such as infectious agents. Sometimes antibodies are produced that bind to substances within the body, resulting in auto-immune disease.

aphantasia The absence of the ability to conjure up mental imagery, the loss of the 'mind's eye'.

aura The prelude to a seizure or migraine attack, usually but not always manifesting as visual disturbance. The symptoms relate to abnormal electrical activity in a specific area of the brain, typically the occipital cortex.

basilar membrane The membrane extending throughout the cochlea, the vibration of which is fundamental to the process of hearing.

blindsight The ability to respond to objects in the field of vision, without conscious perception of those objects.

brainstem The central trunk of the brain, containing all the tracts that flow between the brain and the spinal cord, as well as centres responsible for basic functions like breathing, heart rate, eye movements and swallowing. It consists of the midbrain, the pons and medulla.

cerebellum The region of the brain at the back of the skull, its primary role being in the coordination and regulation of muscle activity and balance.

cerebral cortex The outer layer of the brain, composing of folded grey matter. It plays a fundamental role in consciousness and higher order neurological functions such as cognition, speech and orientation in space. Its precise function relates to its location in the brain.

Charles Bonnet syndrome The experience of visual hallucinations in the presence of eye disease causing visual impairment, with normal neurological function.

cingulate cortex The area of the cerebral cortex deep in the midline, constituting part of the limbic system and thus important for emotional processing, learning and memory. It is highly involved in linking motivation to behaviours.

cochlea The hollow spiral-shaped bone deep in the inner ear, that contains the apparatus to convert the mechanical energy of sound into the electrical energy of nerve impulses.

cornea The transparent layer constituting the front of the eye, overlying the pupil and iris.

cribriform plate The bony structure separating the nasal cavity from the inside of the cranium, it contains multiple perforations through which the nerve fibres responsible for smell run.

deafferentation theory The concept that the removal of inputs to a neurone or network of neurones results in chemical and/or structural changes that promote spontaneous electrical activity in the absence of those inputs.

delusion An abnormal fixed belief held despite being contradicted by rational argument or reality; typically a feature of psychosis.

dorsal columns Tracts within the aspect of the spinal cord closest to the back, responsible for the transmission of sensory information related to light touch, vibration and proprioception, from the body to the brain.

endolymph The fluid that, along with perilymph, fills the hollow structures of the inner ear. The accumulation of endolymph is thought to underlie the development of Ménière's disease.

entorhinal cortex A region of the cerebral cortex that sits deep in the temporal lobe, with an important role in the formation of autobiographical and spatial memories.

fovea The region of the retina which detects visual inputs in the centre of the field of vision, where receptors are highly concentrated to provide maximum resolution.

gustatory Relating to taste.

hallucination The perception of something in the absence of anything present. Like delusions, hallucinations are typical of psychosis, but can also occur in a range of neurological disorders or in normal life. In psychosis, these hallucinations are held to be real.

hippocampus A major component of the temporal lobe, closely linked to the entorhinal cortex, with an important role in the deposition of long-term memories. In Alzheimer's disease, the hippocampus is often an early location of brain shrinkage.

homunculus The representation of the human body within the sensory cortex of the brain; it is distorted so that sensitive parts of the body are hugely over-represented compared to less sensitive areas.

illusion An incorrect or misinterpreted perception of a sensory stimulus. This contrasts with a hallucination, where a sensory stimulus is absent.

limbic system A network of structures primarily concerned with basic emotions such as fear, pleasure and anger, as well as drives such as hunger, thirst or sex. The limbic system also receives a wide array of inputs related to smell, taste and memory.

meningioma A (usually) benign tumour derived from the meninges, the membranous coverings of the brain. These tumours will cause damage through compression or displacement of brain or nerve tissue, although the incidental finding of a meningioma causing no problems at all is common.

mucosa The membranes lining body cavities or organs that are in direct or indirect contact with the outside world, and are kept moist by the presence of mucous glands. These include the linings of the mouth, nose, respiratory system and gut.

multiple sclerosis (MS) A disorder that results in recurrent inflammation of the central nervous system, resulting in relapses characterised by episodes of neurological dysfunction such as visual loss, numbness, weakness or bladder dysfunction. The cause of multiple sclerosis is not fully understood, but it is considered an auto-immune disease.

neurone A nerve cell; the essential building block of the nervous system that transmits electrical impulses.

nociception The process of sensing and perceiving any noxious (potentially damaging) sensory stimuli, including chemical, mechanical or extreme thermal triggers.

occipital Relating to the area at the back of the brain, the occipital lobes, primarily responsible for visual perception and processing.

olfactory Relating to smell.

olfactory bulb The area of the olfactory nerve that sits immediately above the cribriform plate and initially receives smell information from the smell receptors in the nasal mucosa.

olfactory groove The shallow depression in the cribriform plate filled by the olfactory bulb.

olfactory nerve The cranial nerve that runs from the nasal cavity to the brain and is the conduit for all smell information.

optic chiasm The area of the nervous system where the optic nerves meet and cross, resulting in an initial integration of information from both eyes.

optic disc The area of the retina that is the point of entry of the optic nerves. It lacks photoreceptors and therefore represents the physiological blind spot.

optic nerve The cranial nerves that conduct information from each eye towards the brain. The two optic nerves meet at the optic chiasm.

orbitofrontal cortex The area of the cerebral cortex at the front of the brain immediately above the eye sockets. It has a role in decision-making, and in particular expected rewards or punishments related to a particular action. This region is implicated in a range of psychiatric disorders.

ossicles Three tiny bones – the malleus, incus and stapes – that form a chain in the middle ear, transmitting sound impulses from the ear drum into the cochlea.

parietal Relating to the parietal lobe, the area of the brain situated above the occipital lobe toward the crown of the head. Its major function is to process sensory information, particularly with regard to spatial orientation, but it also has a role in attention and mathematical ability.

parosmia The distortion of the sense of smell.

periaqueductal grey A region of the brainstem that acts as a primary control centre for the modulation of pain, as well as having a critical role in a variety of basic functions such as responses to threatening situations.

photoreceptors Specialised nerve cells in the retina that respond to light, generating electrical impulses when stimulated. The two major types are rods, specialised to detect low light levels, and cones, which function at higher light levels but whose pattern of response allows the perception of colour.

primary cortex Refers to a number of different areas of cerebral cortex that are important for the conscious awareness of sensory inputs. These include primary visual, olfactory, auditory and sensory cortices.

proprioception The sensory modality concerned with perceiving the location of the body in space, and of body parts in relation to each other. It is also referred to as joint position sense.

psychosis The presence of hallucinations that are thought to be real and/or delusions, implying a weakening of the grasp of reality. Primarily a feature of psychiatric disease, it can less frequently be seen in neurological disorders.

Riddoch phenomenon The ability to perceive movement in an area of the field of vision that is otherwise blind.

semicircular canals Three small fluid-filled cavities that constitute part of the inner ear. They are orientated in three different axes and detect rotation of the head in all directions.

sensory cortex The area of cerebral cortex whose primary role is the conscious perception of bodily sensations.

sodium channel The molecular apparatus that underlies the conduction of electrical impulses throughout the body. These tiny pores in the outer wall of cells open and close in response to changes in their environment, such as particular chemicals or changes in electrical charge. When they open, sodium channels pass through the pores, causing electrical shifts that can propagate along a cell.

spinothalamic tracts Tracts within the spinal cord that conduct pain and temperature sensations from the body to the brain. In contrast to the dorsal columns, they are located in the part of the cord that is closest to the chest.

stapedius A tiny muscle that dampens the movement of the stapes ossicle, to help protect the ear from loud noises, limiting the conduction of energy into the cochlea.

synapse The small cleft between nerve cells, where impulses are transmitted through the release and detection of chemicals termed neurotransmitters.

synaesthesia The joining or merging of senses that are not usually connected.

temporal Relating to the temporal lobe, the major location for neurological functions such as speech generation and comprehension, visual object recognition and memory.

tinnitus The perception of ringing or buzzing in the ear in the absence of any sound. It represents a form of auditory hallucination.

trigeminal nerve One of the cranial nerves, its main role is the conduction of sensory signals from the face and head. Damage or compression of this nerve may result in severe facial pain, termed trigeminal neuralgia.

Tullio phenomenon The generation of dizziness or vertigo by sounds.

vestibulocochlear nerve The cranial nerve that conducts signals concerning hearing and balance from the inner ear to the brain.

visual cortex The area of the cerebral cortex primarily concerned with visual function, it occupies the surface of the occipital lobe.

ACKNOWLEDGEMENTS

Writers often describe the act of writing as solitary – just one person, a blank sheet of paper or a blinking cursor on a computer screen, filled with white. The only thing between you and an exasperated publisher waiting for the delivery of a manuscript is your own procrastination or writer's block. That may be the case in fiction, but my own experience is incredibly different. I am struck by how many people have had a hand in this book, by the size of the team that has brought it to life.

The stars of these stories are undoubtedly the subjects, those brave individuals who agreed to tell me of their experiences. They are motivated by their hope to alleviate suffering, to act as pathfinders for others who have similar problems and to increase understanding about conditions that most people rarely encounter or think about. These kinds of books that my colleagues and I write are occasionally accused of being like Victorian freak shows, exhibits of the curiosities or horrors of medicine, but this totally misses the point. The telling of these tales serves a greater purpose: to educate, engender empathy and to provide insights into the broader human condition. Apart from one or two cases in which names and other details have been changed to preserve anonymity (and where this happens it is clearly stated), everything remains totally untouched, including direct quotations, unless requested by the speaker, or for clarity. I am enormously grateful to everyone for their active engagement in this book, but particularly so to those who are, or have been, my patients. They have trusted me twice over. It is also important to recognise the work of charities and patient organisations that support individuals out there suffering from these conditions; large, well-known ones like the Royal National Institute of Blind People and the Royal National Institute for the Deaf, but also countless smaller groups, like Esme's Umbrella for Charles Bonnet syndrome, the British Tinnitus Association, and Fifth Sense, a charity supporting people with smell and taste disorders.

As with my first book *The Nocturnal Brain*, the birth of this book and the radio series for the BBC World Service and BBC Radio 4 were inextricably linked. And like *The Nocturnal Brain, The Man Who Tasted Words* owes a significant debt to Sally Abrahams, my brilliant producer at BBC Radio Current Affairs, and to her colleagues Richard Vadon (my executive producer) and Hugh Levinson. It has been a privilege to work with such clever, creative and interested people. Within the context of the radio series, and the experts who gave so much time for background interviews, there are many people to also thank. There are my colleagues at Guy's and St Thomas' and at King's College London: Sui Wong, Dominic Ffytche and Louisa Murdin. Also, from further afield, David Bennett in Oxford, Adam Zeman in Exeter, Julia Simner in Sussex, Carl Philpott in East Anglia, Thomas Hummel in Dresden, Matthew Kiernan in Sydney, Jan Schnupp in Hong Kong, Dana Small at Yale and Donald D. Hoffman at University of California, Irvine – all gave generously of their time, and a few also helped find some of the subjects of this book. Jan Schnupp's excellent book, *Auditory Neuroscience*, co-written with Israel Nelken and Andrew King, was a fantastic reference point for understanding hearing. Donald Hoffman's book, *The Case Against Reality*, blew my mind, and I am still not sure I entirely understand it! It has been a great privilege to extend my learning into realms that lie beyond the traditional limits of clinical neurology, and I hope you will all excuse my more basic understanding of your research and clinical work. I am also grateful to my colleagues at Guy's and St Thomas', and the NHS hospital trust itself, who have humoured my somewhat atypical activities in parallel with my clinical practice.

There are so many other people who have also been fundamental to this book: Luigi Bonomi, my agent in London, without whom I would never have written a word, and his colleagues George Lucas at Inkwell and Nicki Kennedy at ILA. My editors Fritha Saunders at Simon & Schuster in London, and Michael Flamini at St Martin's Press in New York, whose wisdom has refined these chapters, and who commissioned the book in the first place.

I am also grateful to several friends, whose kind critiques have proven enormously helpful. It is always a slightly tentative step, sending out a bit of yourself into the world to people who you have known for almost a lifetime, especially in first draft. So, thanks to Jonathan Turner, Tracey Pettengill Turner and Helen Clarkson, who between them cover all the bases of literature, history and science. My father, Michael Leschziner, approached the manuscript with a diligence usually reserved for peer reviews of scientific articles. I hope he will forgive me for using incomplete sentences, not befitting the scientific journals!

Finally, I cannot be more grateful to my beautiful family. Ava has kept me entertained and distracted over this most horrible year of lockdowns, overflowing hospitals and armchair epide-miology. Maya has brought her dazzling smile, and has even read some of these chapters, ensuring that to some extent it is intelligible. And to my wife, Kavita, without whom none of this would have been possible. She has contributed to every aspect of this book – as a sounding board to bounce ideas off and a fountain of creativity when I have hit a brick wall, passing no comment when I crept off for an hour or two to write. She has been my absolute foundation, for the last year and the preced-ing twenty-four. I am a lucky man, indeed.

FURTHER READING

Much of the science behind the cases in this book can be found in standard textbooks on neuroscience and clinical neurology. There are countless examples, but these are a few of the staples from my shelves:

Kandel, E. R., Schwartz, J. H., Jessell, T. M., Siegelbaum, S. A., Hudspeth, A. J. (eds), *Principles of Neural Science* (5th edition, McGraw-Hill Education, 2012).
Patten, J., *Neurological Differential Diagnosis* (2nd Edition, Springer, 1998).
Clarke, C., Howard, R., Rossor, M. and Shorvon, S. (eds), *Neurology: A Queen Square Textbook* (Wiley-Blackwell, 2009).
Brazis, P. W., Masdeu, J. C., Biller, J., *Localization in Clinical Neurology* (5th edition, Lippincott Williams and Wilkins, 2007).

For more specialised reading, I have included further references below.

Vision

Cowey, A., 'The blindsight saga', *Experimental Brain Research*, 2010, 200(2): 3–24.

Chabanat, E., Jacquin-Courtois, S., Have, L., et al., 'Can you guess the colour of this moving object? A dissociation between colour and motion in blindsight', *Neuropsychologia*, 2019, 128: 204–208.

Ajina, S., Bridge, H., 'Blindsight and unconscious vision: what they teach us about the human visual system', *Neuroscientist*, 2017, 23(5): 529–41.

O'Brien, J., Taylor, J. P., Ballard, C., et al., 'Visual hallucinations in neurological and ophthalmological disease: pathophysiology and management', *Journal of Neurology, Neurosurgery and Psychiatry*, 2020, 91(5): 512–19.

Ffytche, D. H., 'Visual hallucinations in eye disease', *Current Opinion in Neurology*, 2009, 22(1): 28–35.

Leroy R., 'The syndrome of Lilliputian hallucinations', *Journal of Nervous and Mental Disease*, 1922, 56: 325–33.

Teufel, C., Fletcher, P. C., 'Forms of prediction in the nervous system', *Nature Reviews: Neuroscience*, 2020, 21(4): 231–42.

Corlett, P. R., Horga, G., Fletcher, P. C., et al., 'Hallucinations and strong priors', *Trends in Cognitive Sciences*, 2019, 23(2): 114–27.

Adcock, J. E., Panayiotopoulos, C. P., 'Occipital lobe seizures and epilepsies', *Journal of Clinical Neurophysiology*, 2012, 29(5): 397–407.

Hearing and Balance

Schnupp, J., Nelken, I., King, A., *Auditory Neuroscience* (MIT Press, 2012).

Ward, B. K., Carey, J. P., Minor, L. B., 'Superior Canal Dehiscence Syndrome: lessons from the first 20 years', *Frontiers in Neurology*, 2017, 8: 177.

Nakashima, T., Pyykko, I., Arroll, M. A., et al., 'Ménière's disease', *Nature Reviews: Disease Primers*, 2016, 2: 1–18.

Baloh, R. W., 'Prosper Ménière and his disease', *Archives of Neurology*, 2001, 58(7): 1151–6.

Coebergh, J. A. F., Lauw, R. F., Sommer, I. E. C., Blom, J. D. 'Musical hallucinations and their relation with epilepsy', *Journal of Neurology*, 2019, 266(6): 1501–15.

Coebergh, J. A. F., Lauw, R. F., Bots, R., et al., 'Musical hallucinations: review of treatment effects', *Frontiers in Psychology*, 2015, 6: 814.

Baguley, D., McFerran, D., Hall, D., 'Tinnitus', *Lancet*, 2013, 382(9904): 1600–07.

Slade, K., Plack, C. J., Nuttall, H. E., 'The effects of age-related hearing loss on the brain and cognitive function', *Trends in Neuroscience*, 2020, 43(10): 810–21.

Eckert, M. A., Harris, K. C., Lang, H., et al., 'Translational and interdisciplinary insights into presbyacusis: a multidimensional disease', *Hearing Research*, 2021, 402: 108–109.

Corlett, P. R., Honey, G. D., Pletcher, P. D., 'Prediction error, ketamine and psychosis: an updated model', *Journal of Psychopharmacology*, 2016, 30(11): 1145–55.

Touch

Beecher, H. K., 'Pain in men wounded in battle', *Annals of Surgery*, 1946, 123(1): 96–105.

Linden, D. J., *Touch: The Science of the Sense That Makes Us Human* (Penguin, 2015).

Tang, Z., Chen, Z., Tang, B., Jiang, H., 'Primary erythromelalgia: a review', *Orphanet Journal of Rare Diseases*, 2015, 10: 127.

Bennett, D. L. H., Woods, C. G., 'Painful and painless channelopathies', *Lancet: Neurology*, 2014, 13(6): 587–99.

Dib-Hajj, S. D., Waxman, S. G., 'Sodium channels in human pain disorders: genetics and pharmacogenomics', *Annual Review of Neuroscience*, 2019, 42: 87–106.

Isbister, G. K., Kiernan, M. C., 'Neurotoxic marine poisoning', *Lancet: Neurology*, 2005, 4(4): 219–28.

Friedman, M. A., Fleming, L. E., Fernandez, M., et al., 'Ciguatera fish poisoning: treatment, prevention and management', *Marine Drugs*, 2008, 6(3): 456–79.

Shah, S., Vazquez do Campo, R., Kumar, N., et al., 'Paraneoplastic myeloneuropathies: clinical, oncologic, and serologic accompaniments', *Neurology*, 2021, 96(4): e632–9.

Peirs, C., Seal, R. P., 'Neural circuits for pain: recent advances and current views', *Science*, 2016, 354(6312): 578–84.

Besson, J. M., 'The neurobiology of pain', *Lancet*, 1999, 353(9164): 1610–15.

Damien, J., Colloca, L., Bellei-Rodriguez, C.-E., Marchand, S., 'Pain modulation: from conditioned pain modulation to placebo and nocebo effects in experimental and clinical pain', *International Review of Neurobiology*, 2018, 139: 255–96.

Silas Weir Marshall. 'The case of George Dedlow', *Atlantic Monthly*, July 1866, pp.1–10.

Taste and Smell

de Araujo, I. E., Schatzker, M., Small, D. M., 'Rethinking food reward', *Annual Review of Psychology*, 2020, 71: 139–64.

Blankenship, M. L., Grigorova, M., Katz, D. B., Maier, J. X.,

'Retronasal odor perception requires taste cortex, but orthonasal does not', *Current Biology*, 2019, 29(1): 62–9.

Small, D. M., 'Flavor is in the brain', *Physiology and Behavior*, 2012, 107(4): 540–52.

Daniels, J. K., Vermetten, E., 'Odor-induced recall of emotional memories in PTSD – review and new paradigm for research', *Experimental Neurology*, 2016, 284(Pt B): 168–80.

DeVere, R., 'Disorders of Taste and Smell', *Continuum*, 2017, 23(2): 421–46.

Calvi, E., Quassolo, U., Massaia, M., et al., 'The scent of emotions: a systematic review of human intra- and inter-specific chemical communication of emotions', *Brain and Behavior*, 2020, 10(5): e01585.

Genva, M., Kemene, T., K., Deleu, M., et al., 'Is it possible to predict the odor of a molecule on the basis of its structure?', *International Journal of Molecular Sciences*, 2019, 20(12): 3018.

Roper, S. D., Chaudhari, N., 'Taste buds: cells, signals and synapses', *Nature Reviews: Neuroscience*, 2017, 18(8): 485–97.

Kinnamon, S., Finger, T. E., 'Recent advances in taste transduction and signaling', *F1000 Research*, 2019, 8: 2117.

Lübke, K. T., Pause, B. M., 'Always follow your nose: the functional significance of social chemosignals in human reproduction and survival', *Hormones and Behavior*, 2015: 134–44.

Dibattista, M., Pifferi, S., Menini, A., Reisert, J., 'Alzheimer's disease: what can we learn from the peripheral olfactory system?', *Neuroscience*, 2020, 14: 440.

Glezer, A., Bruni-Cardoso, A., Schechtman, D., Malnic, B., 'Viral infection and smell loss: the case of Covid-19', *Journal of Neurochemistry*, 2020, 157(4): 930–43.

Wang, F., Wu, X., Gao, J., et al., 'The relationship of olfactory function and clinical traits in major depressive disorder', *Behavioral Brain Research*, 2020, 386: 112594.

Rochet, M., El-Hage, W., Richa, S., et al., 'Depression, olfaction, and quality of life: a mutual relationship', *Brain Sciences*, 2018, 8(5): 80.

Brennan, P. A., '50 years of decoding olfaction', *Brain and Neuroscience Advances*, 2018, 2: 2398212818817496.

Synaesthesia, Aphantasia and Broader Perception

Hoffman, D. D., *The Case Against Reality: Why Evolution Hid the Truth from our Eyes* (Allen Lane, 2019).

Simner, J., 'Defining synaesthesia', *British Journal of Psychology*, 2012, 103(1): 1–15.

Neckar, M., Bob, P., 'Neuroscience of synesthesia and cross-modal associations', *Reviews in the Neurosciences*, 2014, 25(6): 833–40.

Jewanski, J., Simner, J., Day, S. A., Rothen, N., Ward, J., 'The evolution of the concept of synesthesia in the nineteenth century as revealed through the history of its name', *Journal of the History of the Neurosciences*, 2020, 29(3): 259–85.

Fulford, J., Milton, F., Salas, D., et al., 'The neural correlates of visual imagery vividness – an fMRI study and literature review', *Cortex*, 2018, 105: 26–40.

Zeman, A., Milton, F., Della Sala, S., et al., 'Phantasia – the psychological significance of lifelong visual imagery vividness extremes', *Cortex*, 2020, 130: 426–40.

Zeman, A., Dewar, M., Della Sala, S., 'Lives without imagery – congenital aphantasia', *Cortex*, 2015, 73: 378–80.

Zeman, A., Della Sala, S., Torrens, L. A., et al., 'Loss of imagery phenomenology with intact visuo-spatial task performance: a case of "blind imagination"', *Neuropsychologia*, 2010, 48(1): 145–55.

INDEX